AF598099

SPECIAL TOPICS IN RENEWABLE ENERGY SYSTEMS

Edited by **Ebubekir Yüksel, Abdülkerim Gök** and **Murat Eyvaz**

Special Topics in Renewable Energy Systems
http://dx.doi.org/10.5772/intechopen.73635
Edited by Ebubekir Yüksel, Abdülkerim Gök and Murat Eyvaz

Contributors

Nguyen The Vinh, Vo Vinh, Lorena Amaya-Delgado, Melchor Arellano-Plaza, Anne Gschaedler, Dania Sandoval-Nuñez, Guillermo Flores-Cosio, Javier Arrizon, Omonayo Bolufawi, Mulualem Tesfaye Yeshalem, Baseem Khan, Jorge A. Mejia-Barajas, Mariana Alvarez-Navarrete, Alfredo Saavedra-Molina, Jesús Campos-García, Uri Valenzuela

First published in London, United Kingdom, 2018 by IntechOpen
IntechOpen is the global imprint of INTECHOPEN LIMITED, registered in England and Wales, registration number: 11086078, The Shard, 25th floor, 32 London Bridge Street
London, SE19SG – United Kingdom
Printed in Croatia

British Library Cataloguing-in-Publication Data
A catalogue record for this book is available from the British Library

Additional hard copies can be obtained from orders@intechopen.com

Special Topics in Renewable Energy Systems, Edited by Ebubekir Yüksel, Abdülkerim Gök and Murat Eyvaz
p. cm.
Print ISBN 978-1-78923-979-9
Online ISBN 978-1-78923-980-5

For EU product safety concerns:
IN TECH d.o.o., Prolaz Marije Krucifikse Kozulić 3, 51000 Rijeka, Croatia,
info@intechopen.com or visit our website at intechopen.com.

Meet the editors

Prof. Ebubekir Yüksel is a faculty member of the Environmental Engineering Department at the Gebze Technical University. He received his bachelor degree in Civil Engineering from İstanbul Technical University in 1992. He completed his graduate work (MSc, 1995, and PhD, 2001) at the İstanbul Technical University. His research interests are applications in water and wastewater treatment facilities, electrochemical treatment process and filtration systems at the laboratory, and pilot scale, watershed management, flood control, deep sea discharges, membrane processes, spectrophotometric analyses, chromatographic analyses, and geographic information systems. He has coauthored numerous journal articles and conference papers and has taken part in many national projects. He has produced more than 25 peer-reviewed publications in journals indexed in SCI, SCI-E, and other indexes.

Dr. Abdülkerim Gök is a research associate at Gebze Technical University, Gebze, Kocaeli, Turkey. He completed his bachelor's degree in Materials Science and Engineering at Anadolu University, Turkey, in 2007. He received his MSc degree in Chemical Engineering from Columbia University, New York, USA, in 2011, and his PhD degree in Materials Science and Engineering from Case Western Reserve University, Cleveland, OH, USA, in 2016. His work focuses on developing predictive and mechanistic degradation pathway models of polymeric materials used in photovoltaic module materials under accelerated and natural real-world weathering exposures. His research interests include lifetime and degradation science of PV module and module materials, reproducible statistical methods, and the effect of environmental stressors on PV module performance.

Dr. Murat Eyvaz is an assistant professor of the Environmental Engineering Department at the Gebze Technical University. He received his BSc degree in Environmental Engineering from Kocaeli University in Turkey in 2004. He completed his MSc and PhD in 2013 at the Gebze Institute of Technology (former name of GTU) in Environmental Engineering. He completed his postdoctoral research at the National Research Center on Membrane Technologies in 2015. His research interests are water and wastewater treatment, electrochemical processes, filtration systems/membrane processes, and spectrophotometric and chromatographic analyses. He has coauthored numerous journal articles and conference papers and has taken part in many national projects. He serves as an editor in 30 journals and a reviewer in 100 journals indexed in SCI, SCI-E, and other indexes.

Contents

Preface

In today's world, increased energy needs and environmental and health concerns associated with traditional energy systems have made way for rapid progress in producing energy from renewable resources. Renewable energy is considered to be the answer to future energy demand. Renewable energy is a type of energy that occurs in a natural manner and utilizes unlimited resources, which is the solution for reducing the dependence on fossil fuels and diminishing greenhouse gas emission. It is the key for cleaner, greener, and sustainable energy.

The use of renewable energy systems, with its strong market growth and rapidly dropping prices in recent years, is becoming an increasingly important contributor to the global energy market. Despite the promising potential, wide-scale adoption of known technologies and new alternatives are still essential to address growing world energy needs and compete effectively with fossil fuels and thus diminish greenhouse carbon gas emissions.

This book offers special topics in renewable energy systems such as biofuel generation and adoption of various grid and storage alternatives.

Prof. Ebubekir Yüksel and Asst. Prof. Murat Eyvaz
Department of Environmental Engineering
Gebze Technical University, Turkey

Dr. Abdülkerim Gök
Department of Materials Science and Engineering
Gebze Technical University, Turkey

Chapter 1

Comparative of Lignocellulosic Ethanol Production by *Kluyveromyces marxianus* and *Saccharomyces cerevisiae*

Lorena Amaya-Delgado, Guillermo Flores-Cosío, Dania Sandoval-Nuñez, Melchor Arellano-Plaza, Javier Arrizon and Anne Gschaedler

Additional information is available at the end of the chapter

http://dx.doi.org/10.5772/intechopen.78685

Abstract

The world faces a progressive depletion of its energy resources, mainly fossil fuels based on non-renewable resources. At the same time, the consumption of energy grows at high rates, and the intensive use of fossil fuels has led to an increase in the generation of gaseous pollutants released into the atmosphere, which has caused changes in the global climate. The lignocellulosic bioethanol is considered as a promising alternative for use as fuel ethanol. However, one of the main problems in producing ethanol is toxic compounds generated during hydrolysis of lignocellulosic wastes; these compounds cause a longer lag phase and irreversible cell damage to the microorganisms used in the fermentation step. These conditions of fermentation affect the productivity and the economic feasibility of the lignocellulosic ethanol production process. In this context, many efforts had been carried out to improve the capacity of volumetric ethanol productivity of the yeast. The yeast *Saccharomyces cerevisiae* is commonly employed in industrial ethanol production. However non-*Saccharomyces* yeast as *Kluyveromyces marxianus* can produce alcohols at similar or higher levels than *S. cerevisiae* and on inhibitory conditions.

Keywords: *Kluyveromyces marxianus*, *Saccharomyces cerevisiae*, ethanol, productivity, physiology

1. Introduction

Petroleum-based fuels have been widely used by the human being in daily life and industry for hundreds of years; this has generated a depletion of fossil fuel reserves. In addition, the burning of fossil fuels has caused global climate changes (such as global warming). These concerns have led to the search and development of alternative fuels that are environmentally friendly, renewable, and sustainable. Biofuels such as bioethanol are considered an excellent alternative to mitigate these problems because the production of bioethanol uses renewable resources (such as lignocellulosic biomass) as a raw material; this reduces the dependence on fossil resources and produces cleaner combustion that contributes in a positive way to the environment. Lignocellulosic biomass, in the form of agro-industrial waste, has enormous potential as source energy, precisely as a precursor to bioethanol. The problem of agro-industrial waste in the ethanol production, the potential advantages for the environment, the main stages of production of bioethanol, and a comparison of different microorganisms (yeasts) used for their possible production at the industrial level will be discussed in this chapter. In addition, other bioalcohols that are generated during the production of bioethanol that can be used as high value-added products will be shown.

2. Impact of agro-industrial waste in the ethanol production

The industry has become an essential part of human life; however, development activities produce much waste throughout the world. Large-scale production of agriculture has led to the release of huge quantities of residues. Industries in the agricultural sector generate a large number of residues (in the form of solids, liquids, and gases) throughout the year. Those residues can be used as animal feed, burned (causing an increase in air pollution), or deposited in landfills. The generation of these residues can result in several environmental problems and may cause contamination of air (generating CO_2), contamination of surface water (groundwater seepage). Their elimination is a problem for the producing industries. Therefore, the recovery and re-utilization (recycling) of agro-industrial waste is of interest to industries. This large volume of residues, together with their slow degradation capacity, has stimulated research activities focused on the determination of the possible uses of this waste with a dual purpose. On the one hand, the elimination of an environmental problem (when the waste is discharged) causes the deterioration of the environment since it contains potential toxic compounds, and on the other, the search to produce other value-added products.

Agro-industrial wastes are generated from crop residues or animal feed and include materials such as cereals (corn, rice, wheat, sorghum), bagasses (cane, agave), wheat straw, husk, leaves, seed, stem, and others. Millions of tons of agricultural waste are generated all over the world. **Table 1** shows the generation of agro-industrial wastes according to the region across the world. Rice straw/husk, wheat straw/husk, and sugarcane bagasse are the biggest agro-industrial wastes produced. Brazil, China, India, and the United States account for 60% of the crop residues produced [3]. The agricultural production of maize dominates in the

Residue	Africa	America	Asia	Europe	Oceania	Subtotal	Reference
Crops/grains							
Maize	63.58	445.34	245.75	85.10	0.53	840.3	[1]
Oats	0.20	5.08	0.98	11.95	19.62	37.83	[1]
Rice	34.49	55.49	948.30	6.49	0.31	1045.08	[1]
Rye	0.10	3.77	2.54	34.50	0.81	41.72	[1]
Sorghum	31.66	33.76	14.69	1.06	2.40	83.57	[1]
Sugar cane	22.42	241.15	156.15	0.00	8.39	428.11	[1]
Wheat	33.18	169.19	439.04	305.75	33.89	981.05	[1]
Wheat straw	5.34	62.64	145.20	132.59	8.57	354.34	[1]
Lignocellulosic biomass							
Bagasse	11.73	87.62	74.88	0.01	6.49	180.73	[2]
Corn straw	0.00	140.86	33.90	28.61	0.24	203.61	[2]
Oat straw	0.00	3.04	0.27	6.83	0.47	10.61	[2]
Rice straw	20.9	37.2	667.6	3.9	1.7	731.3	[2]
Sorghum straw	0.00	0.00	9.67	0.35	0.32	10.34	[2]

Table 1. Quantities of potential agro-industrial waste (million tons) available for bioethanol production.

United States and also in China (in less quantity); rice production is mainly found in China and India; cereal production such as wheat is dominant in Europe (outside), China, and India (as they are the major producers of wheat and rice in the world); sugar cane production takes place in Brazil (as major producer) and India. The large portion of agricultural residues generated every year showed that maize has a residue potential of more than 900 million tons of waste; wheat and rice have a residue potential of more than 600 and 400 million tons of waste, respectively. Sugar cane and soybean potentials are in a range of 450 and 350 million tons of waste, respectively [3]. Other studies have also reported that the global production of rice straw is 600–900 million tons per year [4].

The agro-industrial waste contains part of lignocellulosic material; this material is composed by cellulose, hemicellulose, and lignin along with smaller amounts of pectin, proteins, and ashes [5]. Cellulose is the main structural component of the cell wall of plants, and it is in an organized fibrous structure (crystalline), constituting 30–50% of the cell wall. This linear polymer is composed of D-glucose subunits linked together by β 1–4 glycosidic bonds forming cellobiose molecules. These long chains (called elementary fibrils) are linked together by hydrogen bonds and Van der Waals forces that cause cellulose to pack into microfibrils. Hemicellulose and lignin cover the microfibrils forming a matrix. Hemicellulose is a heteropolysaccharide that covers the surface of cellulose fibers and contributes 10–40% of the biomass of the plant, has an irregular structure, and is chemically bound to the lignin in the cell wall. The main characteristic that differentiates hemicellulose from cellulose is that hemicellulose has branches with short side chains that consist of different sugars. These monosaccharides include pentoses (xylose and arabinose), hexoses (glucose, mannose, and galactose), and uronic acids (4-O-methyl-glucuronic, D-glucuronic acid, and D-galacturonic acid) and are linked by β 1–4 glucosidic bonds and occasionally by β 1–3 bonds [6]. Lignin provides rigidity to the cell wall of plants and contributes 15–30% of the biomass of the plant. It is an aromatic polymer of three-dimensional structure quite complex, very branched, and amorphous. Three phenyl propionic alcohols exist as monomers of lignin: coniferyl alcohol, p-coumaryl alcohol, and sinapyl alcohol. Alkyl-aryl, alkyl-alkyl, and aryl-aryl ether linkages hold these phenolic monomers together.

The composition of these components can vary from one plant species to another. For example, wood has a higher amount of cellulose, while straw and wheat leaves increase the content of hemicellulose [7]. Besides, the relationships between the various components within a single plant vary with age, growth stage, and other culture conditions [6]. Practically, all biomass residues produced in agricultural and industrial activities, and even urban waste, have high concentrations of available lignocellulosic materials. Lignocellulosic materials represent the most abundant renewable resources on earth and therefore have been a great deal of interest in utilizing as energy resource. In the last decades, the bioconversion of lignocellulosic materials in products of commercial interest (biofuels, bioalcohols) has been searched. Agro-industrial wastes, which are byproducts of key industrial and economical activities, are attractive raw materials for the production of renewable fuels, like bioethanol. The great advantage of using these materials is that they are natural, biodegradable, and can often be extracted from waste or as byproducts of the agricultural or food industry. The advantage associated with these products is double, since the cost of the raw material becomes cheap and, also, an added value is given to the industrial waste or byproduct. The use of lignocellulosic biomass as a raw material for the production of biofuels (such as bioethanol) has been associated with a concept of biorefinery, which is key to the production of ethanol at an industrial level. However, the bioconversion of lignocellulosic wastes into useful products is an enormous environmental challenge.

3. Lignocellulosic ethanol

Currently, the world faces a progressive depletion of its energy resources, mainly fossil fuels based on non-renewable resources. At the same time, the consumption of energy grows at high rates and the intensive use of fossil fuels has increased the generation of gaseous pollutants released into the atmosphere, which has caused changes in the global climate. Through the use of renewable energy resources, it is possible to find part of the solution to the energy requirements in a friendly way for the environment. The global potential of bioenergy is represented in energy crops and lignocellulosic waste [1, 8]. The conversion of these raw materials into biofuels is an option for the exploitation of alternative sources of energy and the reduction of polluting gases [9]. In addition, the use of biofuels has important economic and social effects.

Bioenergy companies focus on first-generation biofuels and involve the use of grains and raw materials directly related to human consumption (grains, sugarcane, etc.). The use of these raw materials implies that the processes to produce first-generation fuels are not sustainable and at the same time are considered not ethic (it comes from feedstocks directly related to human or animal feed). On the other hand, second-generation biofuels are those that use non-food raw materials such as waste from agro-industries (straw, bagasse, husks, effluents). The second-generation biofuels market is based on the basic principle that the majority will be produced from agro-industrial waste, which implies the creation of sustainable developments and clean and friendly companies with the environment. Governments envision the second-generation biofuels market as an innovative way to reduce costs in the disposal of waste materials and improve the production of clean, renewable, and sustainable energies, which in the long term partially replace the use of fossil fuels such as petroleum.

Nowadays, bioethanol is the most attractive and important renewable fuel in terms of capacity and market value [10, 11]. Globally, bioethanol production has increased considerably in recent years and has gained importance in many parts of the world. The largest producer of bioethanol is located in America, and Asia stands second while Europe follows the list (**Figure 1**). Several reports have indicated the world production capacity of bioethanol; in 2012 and 2013, it was approximately 234 billion liters per year [12]. More than 128.5 billion liters of bioethanol were produced worldwide during 2014 [13], while for 2016, it increased to almost 144 billion liters [14]. Brazil and the United States are the main producers of first-generation bioethanol (which represent around 60% of world production) (**Figure 1**) [2]. In the United States, ethanol production increased from 175 million gallons to 15,800 million gallons in almost 20 years [12]. For 2017, the annual production of bioethanol was 27,050 million liters (the United States, Brazil, Europe, and China being the main contributors).

The production of second-generation bioethanol is shown in **Table 2**. The production of bioethanol from agro-industrial waste has taken much interest about the first generation, due,

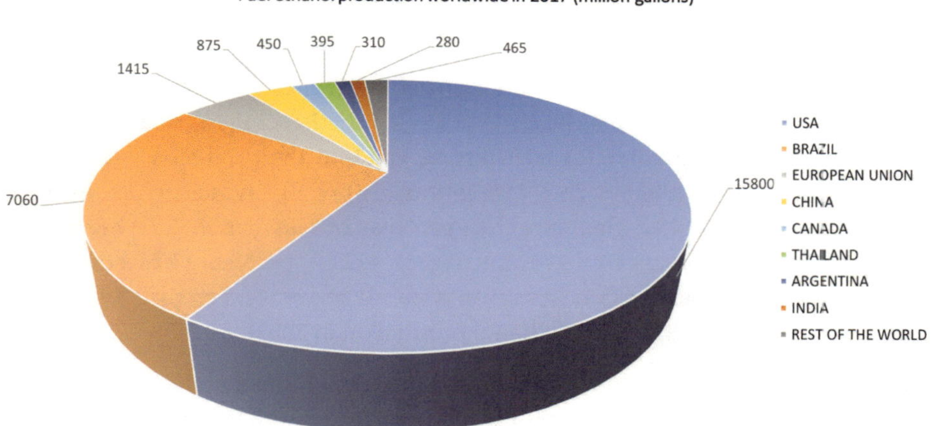

Figure 1. Worldwide fuel ethanol production in 2017.

Residue	Africa	America	Asia	Europe	Oceania	Subtotal
Waste crop						
Barley	0.12	0.045	0.83	1.35	0.13	2.47
Corn	2.17	4.29	6.82	1.09	0.01	14.38
Oat	0.02	0.044	0.04	0.30	0.001	0.405
Rice	0.71	1.61	14.4	0.02	0.02	16.76
Sorghum	1.55	0.21	0.37	0.003	0.0004	2.13
Sugar cane	0.23	0.55	0.82	—	0.0001	1.60
Wheat	0.55	0.78	6.78	2.70	0.54	11.35
Subtotal (1)	5.35	7.529	30.06	5.463	0.7015	49.103
Lignocellulosic biomass						
Bagasse	3.33	24.87	21.3	0.0004	1.84	51.34
Barley straw	—	3.2	0.61	13.7	0.60	18.11
Corn stover	—	40.47	9.75	8.23	0.07	58.52
Oat straw	—	0.799	0.07	1.79	0.12	2.77
Rice straw	5.86	10.41	186.8	1.10	0.47	204.64
Sorghum straw	—	2.61	—	0.10	.09	2.8
Wheat straw	1.57	18.39	42.6	38.9	2.51	103.97
Subtotal (2)						
Total (1 + 2)	16.11	108.278	291.19	69.2834	6.4015	491.2535

Table 2. Potential bioethanol production (gigaliters) from agro-industrial wastes by continent [1, 2].

the main feedstock for the production are residues, in that way avoiding the use of extensions of cultivable land to bioethanol production. In addition, as mentioned earlier, a product of added value is obtained from residues which also help to avoid a problem of waste accumulation. The selection of the agro-industrial feedstock is in function of the agricultural production and the interests of each country for transferring value to the produced wastes. Kim and Dale [2] indicated that area about 73.9 Tera grams (Tg) of dry wasted crops in the world could potentially produce 49.1 gigaliters (GL) per year of bioethanol. Also mentioned was that the lignocellulosic biomass could produce up to 442 GL per year of bioethanol and the total potential bioethanol production from crop residues and wasted crops is 491 GL per year (about 16 times higher than the world ethanol production in 2003).

Bioethanol presents some important differences about conventional fuels derived from petroleum. The main one is the high oxygen content, which constitutes about 35% by mass of ethanol. The characteristics of bioethanol allow a cleaner combustion (it emits low levels of non-combusted hydrocarbons, such as carbon monoxide (CO), oxides of nitrogen (NOx), and other reactive organic gases that pollute the air) [12, 15]. Also, it improves the performance of the engines, which contributes to reduced pollutant emissions (with low emission of CO_2 to the atmosphere), even when mixed with gasoline [9, 16]. In this way, bioethanol (and biofuels) helps to reduce greenhouse gas (GHG) emissions and consequently mitigates the climate change. Also, the use of agro-industrial waste for the production of bioethanol helps to reduce their accumulation which is of great environmental concern. Farrell et al. [17] estimated (using the displacement method) a reduction of the 88% greenhouse gas emissions using lignocellulosic ethanol (from switchgrass). Schmer et al. [18] reported that ethanol from switchgrass reduces GHG emissions by 94% compared to GHG emissions from gasoline. Various studies have reported a reduction of 63–118% in life-cycle GHG emissions using ethanol from lignocellulosic feedstock in comparison to fossil fuel [9, 19–24].

The main steps that are involved in the production of bioethanol from lignocellulosic wastes are 1) pre-treatment, 2) hydrolysis and 3) fermentation (**Figure 2**). An adequate process of pre-treatment and hydrolysis can increase the concentrations of fermentable sugars, which potentially would help to obtain greater quantities of bioethanol.

3.1. Pre-treatment

The aim of the pre-treatment is to disintegrate the lignocellulosic complex to make it more accessible for the hydrolysis stage; in this way, it helps decrease the crystallinity of the cellulose, increase the surface area of the biomass, remove the hemicellulose, and break the seal of the lignin. This process increases the porosity of the pretreated material and makes cellulose more accessible to enzymes so that the conversion of carbohydrates into fermentable sugars is achieved quickly and with higher yields. The pre-treatment methods can be divided into different categories: physical (grinding), chemical (alkaline, dilute acids, oxidizing agents, organic solvents), biological, or a combination of these.

3.2. Hydrolysis

After pre-treatment, cellulose and hemicellulose are released and converted to monomers which are known as saccharification (containing mainly glucose and pentose); the reaction can be catalyzed by diluted acids, concentrated acids, or enzymes (cellulases and hemicellulases). Acid hydrolysis (both concentrated and diluted) occurs in two stages, to take advantage of the differences between hemicellulose and cellulose. The first involves, essentially, the hydrolysis of the hemicellulose, conducted in accordance with the pre-treatment conditions discussed earlier. In the second stage, higher temperatures are applied, seeking to optimize the hydrolysis of the cellulose fraction [25]. The process with concentrated acid uses high temperatures and pressures, with reaction times of seconds to a few minutes, which facilitates the use of continuous processes. On the other hand, the processes with diluted acid develop under less severe conditions, but with typically long reaction times [26]. In the enzymatic process, hydrolysis is catalyzed by enzymes generically cellulases and requires at least three key enzymes, endoglucanases, exoglucanases, and β-glucosidases. Enzymatic hydrolysis is a prolonged process because the enzymes are hampered by the structural parameters of the substrate (hemicellulose/lignin) and the surface area. However, enzymatic hydrolysis has certain advantages over acid hydrolysis: the first is that by not using chemicals, it is an ecological

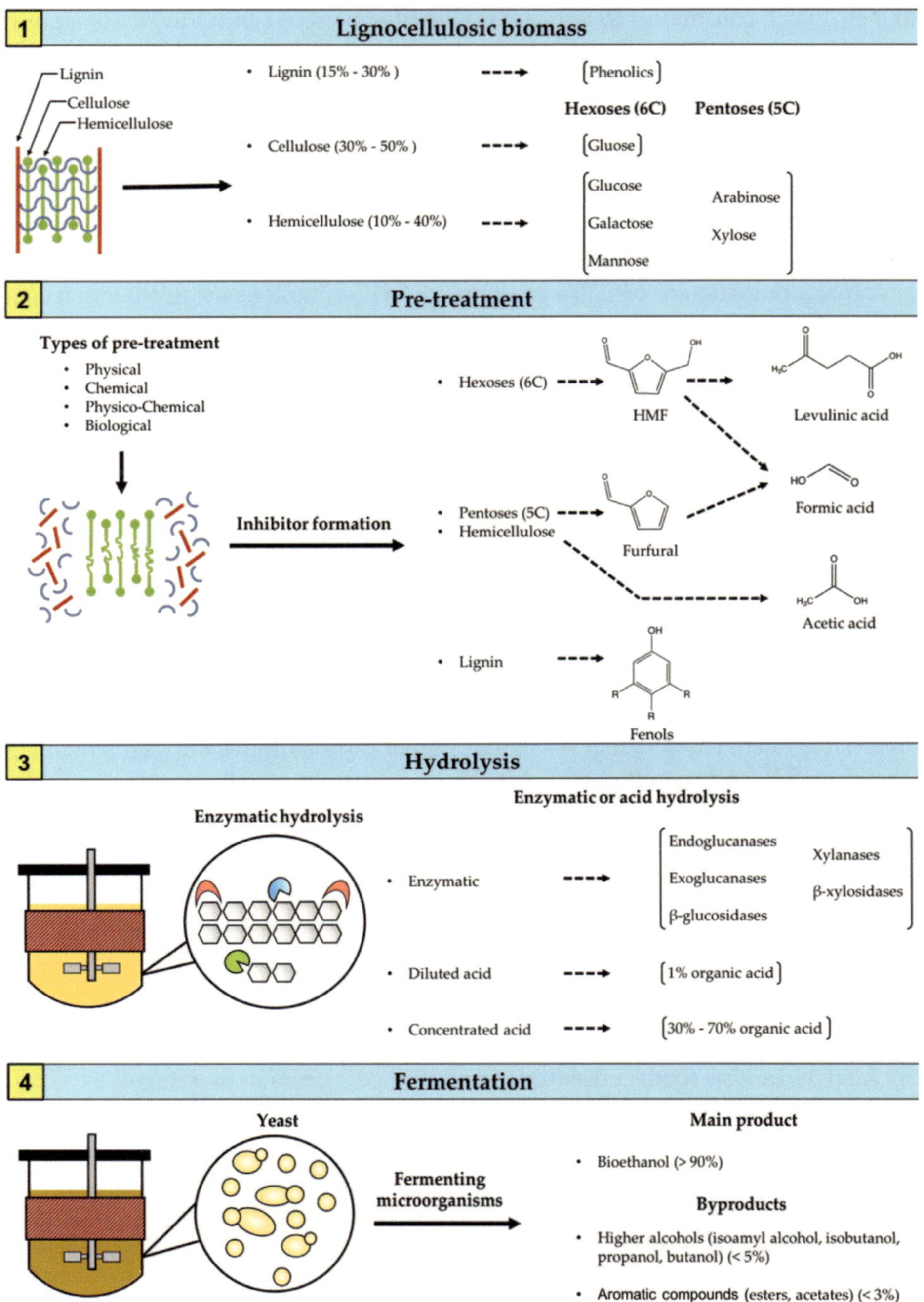

Figure 2. Schematic diagram of bioethanol production from lignocellulosic biomass. 1: Composition of lignocellulosic biomass; 2: Effects of pre-treatment on lignocellulosic biomass and main/representative inhibitory compounds generated during the process. Pre-treatment is required to separate the cellulose, hemicellulose and lignin fractions; 3: Hydrolysis of pre-treated lignocellulosic material. Hydrolysis helps to convert polymeric carbohydrates (cellulose and hemicellulose) into fermentable sugars; 4: Fermentation process. The fermentation process by microorganisms converts the soluble sugars released during hydrolysis into ethanol and byproducts.

alternative, a second is that they are carried out under moderate environmental conditions of temperature and pH, which reduces their cost in comparison with acid hydrolysis; they also avoid corrosion problems, and additionally, toxic byproducts (inhibitors) are not formed.

3.3. Fermentation

After hydrolysis, the product of the sugars is fermented to ethanol by microorganisms. The microorganisms can be either bacteria or yeast that can use one or more of the sugars present in the lignocellulosic material pretreated and hydrolyzed.

The typical configuration used for the fermentation of lignocellulosic hydrolysates involves a sequential process in which hydrolysis (cellulose/hemicellulose) and fermentation are carried out in different units. This configuration is known as separate hydrolysis and fermentation (SHF). SHF has the advantage that each step can be carried out under optimal conditions of hydrolysis (45–50°C) and fermentation (30–35°C). SHF has the disadvantage that enzymes (cellulases/β-glucosidases) are inhibited by the glucose released during hydrolysis; therefore, to obtain optimum yields, it requires a lower charge of solids and the addition of a higher load of enzymes. In addition, due to the fact that the process is carried out in different steps, it has the disadvantage that increases its cost, which compromises its economic viability. An alternative to this process is the known simultaneous saccharification and fermentation (SSF), where hydrolysis and fermentation are carried out in a single unit. SSF considerably reduces cellulase inhibition problems (which improves yields and production bioethanol). Also, during the SSF method, the risk of contamination can be reduced (due to the ethanol generated), and the use of a single unit reduces the costs of the process. However, it has the disadvantage that each stage cannot be optimized (limiting and restricting the process). A variant of SSF is simultaneous saccharification and co-fermentation (SSCF) and non-isothermal simultaneous saccharification and fermentation (NSSF). SSCF consists of the enzymatic hydrolysis and co-fermentation of pentoses and hexoses in a single unit; the co-fermenting microorganisms need to be compatible regarding operating pH and temperature. In NSSF, hydrolysis and fermentation occur simultaneously in two different units at their optimum temperature; the hydrolysis reaction liquid passes to the fermenter. An alternative is consolidated bioprocessing (CBP), also known as direct microbial conversion (DMC). In this method, all the biological transformations involved in the production of bioethanol (enzyme production (cellulase), biomass hydrolysis (saccharification), fermentation of hexoses, and fermentation of pentoses) occur in a single unit with one or more microorganisms. This method is attractive because it reduces the number of units, the cost of the products, and simplifies its operation.

Currently, the tendency of these methods focuses on carrying out the processes simultaneously in a smaller number of steps. The microorganisms are fundamental pieces in the production of lignocellulosic bioethanol. Microorganisms capable of fermenting both sugars with high yield are required. One of the main problems in the production of bioethanol from lignocellulosic hydrolysates is that not all microorganisms can assimilate and ferment pentoses to ethanol; the microorganisms conventionally used in the fermentation of bioethanol present this inconvenient. As mentioned earlier, hemicellulose can represent a large part of the lignocellulosic material, which can lead to a large number of available pentoses for fermentation.

For this reason, the development and search for microorganisms capable of efficiently fermenting pentose to ethanol have led to a great research effort. Genetic engineering has been used to add pentose metabolic pathways in yeast and other microorganisms to effectively utilize the sugars present in lignocellulosic hydrolysates. In addition, several types of research have sought to improve the performance of microorganisms that already can ferment both sugars. Although success has been achieved in this regard, the fermentation of mixtures of the sugars of the lignocellulosic biomass has not yet reached a commercially viable level.

4. Fermentative behavior of *Saccharomyces cerevisiae* and *Kluyveromyces marxianus* in the bioethanol production

S. cerevisiae is the main microorganism employed in ethanol production due to its high ethanol productivity, high ethanol tolerance, and the ability to ferment a wide range of sugars. However, there are some adverse effects in yeast fermentation which inhibit ethanol production such as high temperature, inhibitory compounds from the hydrolysis of lignocellulosic residues (furans, aldehydes, and organic acids), high concentrations of carbon and ethanol source, and the ability to ferment pentose sugars. Different types of yeast strains have been used in the fermentation process for ethanol production (**Table 3**) including hybrid, recombinant, and wild-type yeasts. The production of bioethanol depends on several factors, such as the concentration of sugars, pH, temperature, the concentration of inhibitor compounds, fermentation time, cell growth rate, agitation, and inoculum size [40].

K. marxianus, a non-conventional thermotolerant yeast, is potentially useful for the production of ethanol and other products because they can assimilate diverse sugars including xylose, arabinose, sucrose, raffinose, and inulin in addition to several hexoses [41]. In addition, *K. marxianus* has a faster growth and ethanol production at higher temperatures [42]. Flores et al. [43] reported several strains of *K. marxianus* capable of producing ethanol from *Agave tequilana* fructans (ATF) at high temperatures, which evidence the high potential of *K. marxianus* for the simultaneous saccharification and fermentation for bioethanol production. Other investigations have also reported that *K. marxianus* shows a better behavior to produce ethanol under inhibitory conditions (furans) compared to a commercial strain of *S. cerevisiae* due to the best response and resistance to inhibitors (**Table 4**) [27].

The presence of inhibitors is another problem faced by yeast during the fermentation of lignocellulosic hydrolysates. The presence of inhibitory compounds originated in pre-treatment, and hydrolysis processes of lignocellulosic biomass can strongly inhibit the fermentation stage and generate irreversible cellular damage (which negatively affect yeast metabolism). Furan aldehydes such as 2-furaldehyde (furfural) and 5-hydroxymethylfurfural (HMF) are products of degradation of pentoses and hexoses released from hemicellulose and cellulose, respectively. Acetic, formic, and levulinic acids are the most common acids present in lignocellulosic hydrolysates. Acetic acid is formed by deacetylation of hemicelluloses, while formic

Yeast strain	Type of strain	Waste/medium	Sugar concentration (g L^{-1})	Ethanol concentration (g L^{-1})	Qp	Reference
S. cerevisiae ERD	Commercial	Mineral medium	40	18.2	1.51	[27]
S. cerevisiae	Laboratory	Sugarcane bagasse	30	13.1	0.18	[28]
S. cerevisiae AR5	Laboratory	Sugarcane bagasse	18.5	2.2	0.09	[29]
S. cerevisiae ERD	Laboratory	Sugarcane bagasse	18.5	4.8	0.17	[29]
S. cerevisiae AR5	Laboratory	Wheat straw	14.5	2.4	0.21	[29]
S. cerevisiae ERD	Laboratory	Wheat straw	14.5	2.2	0.20	[29]
S. cerevisiae RL-11	Laboratory	Spend coffee grounds	195.0	11.7	0.49	[30]
S. cerevisiae MTCC 173	Laboratory	Sorghum stover	200.0	68.0	0.94	[31]
S. cerevisiae KL17	Wild-type	Galactose and glucose	500.0	96.9	3.46	[32]
S. cerevisiae CHY1011	Wild-type	Cassava starch	195.0	89.9	1.35	[33]
S. cerevisiae RPRT90	Mutated	Ipomea carnea	72.1	29.0	1.03	[34]
K. marxianus K213	Laboratory	Water hyacinth	23.3	7.34	0.31	[35]
K. marxianus CECT10875	Laboratory	Wheat straw	—	36.2	0.50	[36]
K. marxianus SLP1	Laboratory	Sugarcane bagasse	18.5	9.0	0.29	[29]
K. marxianus OFF1	Laboratory	Sugarcane bagasse	18.5	5.1	0.20	[29]
K. marxianus SLP1	Laboratory	Wheat straw	14.5	3.1	0.25	[29]
K. marxianus OFF1	Laboratory	Wheat straw	14.5	2.3	0.20	[29]
K. marxianus UMIP2234.94	Wild type	Glucose	20.0	5.7	0.23	[37]
K. marxianus 10,875	Laboratory	Wheat straw	—	11.1	0.20	[38]
K. marxianus DBKKUY-103	Laboratory	Sorghum juice	194	85.16	1.42	[39]

Qp, volumetric ethanol productivity (g ethanol produced L^{-1} h^{-1}).

Table 3. Bioethanol production from different carbon sources.

and levulinic acids are products of HMF breakdown/degradation. In addition, formic acid can be formed from furfural under acidic conditions at elevated temperatures. A wide range of phenolic compounds is generated due to the decomposition of lignin and also by the degradation of carbohydrates during acid hydrolysis. Furans are considered the most representative

Stress condition	Lag phase	μ	Rs	Yx/s	qs	Qs	Yp/s	qp	Qp
	ERD								
Control	0	0.14	3.83	0.059	2.47	1.55	0.49	1.21	0.30
HMF (7 g L^{-1})	0	0.07	1.74	0.043	1.67	1.62	0.45	0.75	0.29
Furfural (3 g L^{-1})	6	0.08	1.87	0.047	1.76	1.65	0.46	0.82	0.30
HMF (3.5 g L^{-1}) + furfural (1.5 g L^{-1})	6	0.12	1.52	0.045	2.78	1.65	0.46	1.29	0.30
HMF (7 g L^{-1}) + furfural (3 g L^{-1})	—	—	—	—	—	—	—	—	—
	SLP1								
Control	0	0.16	3.20	0.053	3.03	1.66	0.40	1.22	0.26
HMF (7 g L^{-1})	12	0.10	1.32	0.049	2.11	1.59	0.43	0.91	0.27
Furfural (3 g L^{-1})	6	0.07	1.75	0.048	1.60	1.55	0.48	0.78	0.30
HMF (3.5 g L^{-1}) + furfural (1.5 g L^{-1})	6	0.06	1.25	0.046	3.46	1.48	0.49	1.73	0.29
HMF (7 g L^{-1}) + furfural (3 g L^{-1})	—	—	—	—	—	—	—	—	—

Lag phase (h); μ, specific growth rate (h^{-1}); Rs, substrate consumption rate (g L^{-1} h^{-1}); Yx/s, biomass substrate yield (g dry cell weight g substrate utilized^{-1}); qs, specific substrate consumption rate (g substrate consumed g dry cell weight^{-1} h^{-1}); Qs, volumetric substrate uptake rate (g substrate consumed L^{-1} h^{-1}); Yp/s, ethanol yield on substrate (g ethanol produced g substrate utilized^{-1}); qp, specific ethanol productivity (g ethanol produced g dry cell weight^{-1} h^{-1}); Qp, volumetric ethanol productivity (g ethanol produced L^{-1} h^{-1}).

Table 4. Physiological parameters of ***Saccharomyces cerevisiae*** ethanol red (ERD) and ***Kluyveromyces marxianus*** SLP1 under different inhibitor conditions in mineral medium (at 100 rpm, 30°C, pH 4.5) (taken from [27]).

and inhibitory toxic compounds of the fermentative capacities of yeasts. Furfural significantly reduces cell proliferation, ethanol production, inhibits several enzymes that are essential to central metabolism, including dehydrogenases, or by damaging and blocking the synthesis of DNA, RNA, proteins, carbohydrate metabolism, and changes in the cell wall. On the other hand, HMF damages membranes and nucleic acids, generates oxidative stress, and reduces NADH/NADPH and inhibition enzymes [44, 45].

Furfural and HMF have been studied mostly in *S. cerevisiae* showing decreases in ethanol yield. These compounds affect the physiology of the yeast and often result in a decreased viability, a lower metabolite yield, and a diminished productivity during biofuels generation [45]. *K. marxianus* has also been considered as a host for bioethanol production from lignocellulosic biomass hydrolysates [29]. It has been reported that inhibitors from the lignocellulosic material are related to the levels of reactive oxygen species (ROS). A typical 2-Cys peroxiredoxin from *K. marxianus* Y179 (KmTPX1) was identified, and its over expression was achieved in *S. cerevisiae* 280. Strain TPX1 with overexpressed KmTPX1 gene showed enhanced tolerance to oxidative stresses [46]. Considerable decreases in cell growth and ethanol yields have been observed at high concentrations of furans in *K. marxianus* (**Table 5**) (Sandoval et al., unpublished data).

Furfural (g/L)	Kinetic parameters		
	Yp/s	**Yp/x**	**Qp**
Control	0.42 ± 0.012	0.67 ± 0.006	0.17 ± 0.013
0.5	0.29 ± 0.022	0.63 ± 0.012	0.11 ± 0.063
1	0.24 ± 0.013	0.23 ± 0.030	0.10 ± 0.001
2	NG	NG	NG
3	NG	NG	NG
4	NG	NG	NG
HMF (g/L)	**Yp/s**	**Yp/x**	**Qp**
Control	0.42 ± 0.019	0.22 ± 0.026	0.17 ± 0.016
1	0.18 ± 0.023	0.12 ± 0.048	0.05 ± 0.005
2	0.18 ± 0.035	0.01 ± 0.023	0.05 ± 0.002
4	0.17 ± 0.027	0.09 ± 0.003	0.04 ± 0.013
6	0.15 ± 0.008	0.09 ± 0.004	0.04 ± 0.006
8	0.12 ± 0.003	0.08 ± 0.012	0.04 ± 0.012
10	0.10 ± 0.013	0.06 ± 0.011	0.04 ± 0.007

Mineral medium (glucose 20 g L^{-1}), 30°C, pH 4.5 and 100 rpm. Yp/s, ethanol yield on substrate (g ethanol produced g substrate utilized^{-1}); Yp/x, ethanol yield on biomass (g ethanol produced g dry cell weigh^{-1}); Qp, volumetric ethanol productivity (g ethanol produced L^{-1} h^{-1}); NG, no growth detected.

Table 5. Kinetic parameters of fermentation by ***Kluyveromyces marxianus*** SLP1.

5. Higher alcohols and aromatic compounds production from *S. cerevisiae* and *K. marxianus*

In addition to the production of bioethanol, other byproducts (such as fusel oil) can be generated during the metabolic processes of the yeast in fermentation. Fusel oil is a mixture of higher alcohols (from amino acid metabolism) mainly composed by isoamyl alcohol, isobutanol, propanol, butanol, and other aromatic compounds, such as esters and acetates. Higher alcohols can be used as fuel or energy source (when burned) [47]. Esters can also be obtained from higher alcohols of fuel oil, which are used as solvents, flavoring agents, medicinal, and plasticizers [48]. Isoamyl esters (such as isoamyl acetate) are aromatic compounds that are widely used in the food and beverage industry (because they have important compounds of flavor and fragrance) [49]. Other aromatic compounds produced by yeasts, such as acetates, also show commercial applications and have been used as components in flavorings (ethyl acetate, isobutyl acetate, amyl acetates, and isoamyl acetate) and additives for perfume (isopropyl acetates, acetates of octyl, and methyl acetates). Ethyl acetate is one of the most important esters and can be used as a chemical solvent, resins, adhesives, and paints [50].

Many of the studies link the formation of various aromatic compounds such as fusel alcohols and fusel acids to the degradation of branched-chain and aromatic amino acids, via the Ehrlich pathway [51, 52]. The synthesis of higher alcohols synthesized in the yeast catabolism was proposed by Ehrlich 1907; from this finding, the Ehrlich pathway has been used as a basis for the modification of routes and enzymes involved in the formation of higher alcohols. Ehrlich pathway proceeds in three steps: first, the amino acid is transaminated to create an α-keto acid; next, this α-keto acid is decarboxylated to an aldehyde; and finally, the aldehyde is reduced to an alcohol or oxidized to a fusel acid, depending on the redox status of the cells. In addition to their synthesis via the Ehrlich pathway, α-keto acids are intermediates in the biosynthesis of certain amino acids and therefore constitute a link between anabolic and catabolic processes. Depending on metabolic fluxes, unbalanced amino acid synthesis can also contribute to the synthesis of aromatic molecules [53].

K. marxianus is considered a yeast with high potential for the generation of compounds of industrial value from lignocellulosic waste; in addition to consuming different sugars from waste, it can convert the available sugars into molecules of industrial interest such as higher alcohols and aromatic molecules, mainly acetate esters. **Table 6** shows the production of ethanol and other metabolites at industrial level of some strains of yeast [43, 54]. Other studies at laboratory level have reported that *S. cerevisiae* and *K. marxianus* can produce volatile compounds, even in the presence of furans (**Table 7**) [27].

Usually, in Mexico, the strains of *K. marxianus* are isolated from environments with extreme conditions (such as fermentation process of mezcal or tequila); this allows the yeast to develop specific mechanisms to survive in those hostile environments. Therefore, some strains of *K. marxianus* could be a good candidate to be used at an industrial level to produce lignocellulosic ethanol.

Strain	Ethanol	Higher alcohols	Esters	Aldehydes	Methanol
S. cerevisiae (AR5)*	48.0	26.612	1.742	1.027	31.07
K. marxianus (GRO6)*	54.5	33.769	2.023	0.868	27.907
K. marxianus (DU3)+	47.5	84.68	84.9	25.8	3.8
K. marxianus (SLP1)+	44.9	62.08	112.7	27.3	3.6
K. marxianus (OFF1)+	49.7	101.61	120.7	25.6	4.3
K. marxianus (DH)+	46.0	67.71	101.9	23.7	3.8
K. marxianus (DZ5)+	47.1	81.53	105.3	12.6	3.7
K. marxianus (DA5)+	49.9	75.2	106.3	33.3	3.8
K. marxianus (DL)+	45.9	75.4	118	17.7	4.1

*Hydrolysis and separate fermentation (SHF).

+Simultaneous saccharification and fermentation (SSF).

Table 6. Generation of ethanol (g L^{-1}), higher alcohols (mg L^{-1}), esters (mg L^{-1}), aldehydes (mg L^{-1}), and methanol (mg L^{-1}) at the end of fermentation.

Compound (mg L[-1])	Strain				
	AR5	ERD	MC4	SLP1	OFF1
Aldehydes					
Acetaldehyde	31.42	20.47	30.05	33.59	25.30
Esters					
Ethyl acetate	1.21	1.32	1.24	4.11	4.29
Isoamyl acetate	0.00	0.00	0.00	0.00	0.00
Ethyl lactate	0.65	1.22	1.17	0.59	0.61
Ethyl hexanoate	6.41	7.34	6.77	6.93	8.19
Ethyl octanoate	1.03	0.90	0.89	1.00	1.01
Alcohols					
Ethanol	4910	5120	5300	5270	4340
1-propanol	5.92	5.57	5.54	6.89	6.63
Isobutanol	4.51	5.03	3.58	20.48	16.87
Butanol	5.04	3.32	3.76	1.78	1.43
Amyl alcohol	21.63	19.88	22.06	47.78	41.74
2-phenyl-ethanol	0.34	0.33	0.38	0.32	0.34

Table 7. Ethanol and volatile compounds produced by strains ***S. cerevisiae*** and ***K. marxianus*** in defined medium with glucose (20 g L^{-1}) as the sole carbon source (100 rpm, pH 4.5, 30°C) (taken from [27]).

6. Conclusion

The use of bioethanol as fuel has helped to reduce the consumption of fossil fuels and the problems of pollution worldwide; also, the use of lignocellulosic waste for the production of second-generation bioethanol benefits the environment. However, to produce bioethanol, one of the critical points in the process is fermentation, due to the lack of robust microorganisms that can efficiently convert lignocellulosic hydrolysate sugars to ethanol. In the last years, the yeast *K. marxianus* has been widely tested in diverse biotechnological applications, including the production of bioethanol and has shown that it can match the bioethanol yields of commercial *S. cerevisiae* strains, and in many cases, it can be superior. The capacities of *K. marxianus* to be thermotolerant have a high growth rate, consume different carbon sources, and be resistant to toxic compounds, allowing it to have higher yields and ethanol production than *S. cerevisiae*; therefore is considered that this yeast has a good potential for lignocellulosic bioethanol production at the industrial level.

Acknowledgements

The authors were supported by research fellowships from the National Council of Science and Technology from Mexico and the Mexican Ministry of Energy (FSE CONACYT-SENER 245750, 248090).

Conflict of interest

The authors declare no conflict of interest.

Author details

Lorena Amaya-Delgado*, Guillermo Flores-Cosío, Dania Sandoval-Nuñez, Melchor Arellano-Plaza, Javier Arrizon and Anne Gschaedler

*Address all correspondence to: lamaya@ciatej.mx

Center for Research and Technological Assistance and Design of the State of Jalisco AC (CIATEJ), Guadalajara, Mexico

References

[1] Gupta A, Verma JP. Sustainable bio-ethanol production from agro-residues: A review. Renewable and Sustainable Energy Reviews. 2015;**41**:550-567. DOI: 10.1016/j.rser.2014.08.032

[2] Kim S, Dale BE. Global potential bioethanol production from wasted crops and crop residues. Biomass and Bioenergy. 2004;**26**:361-375. DOI: 10.1016/j.biombioe.2003.08.002

[3] Bentsen NS, Felby C. Technical potentials of biomass for energy services from current agriculture and forestry in selected countries in Europe, the Americas and Asia. Forest and Landscape Working Papers. 2010;**54**:31

[4] Karimi K, Kheradmandinia S, Taherzadeh MJ. Conversion of rice straw to sugars by dilute-acid hydrolysis. Biomass and Bioenergy. 2006;**30**:247-253. DOI: 10.1016/j.biombioe.2005.11.015

[5] Jorgensen H, Kristensen JB, Felby C. Enzymatic conversion of lignocellulose into fermentable sugars: Challenges and opportunities. Biofuels, Bioproducts and Biorefining. 2007;**1**:119-134

[6] Perez J, Munoz-Dorado J, de la Rubia T, Martinez J. Biodegradation and biological treatments of cellulose, hemicellulose and lignin: An overview. International Microbiology. 2002;**5**:53-63

[7] Sun Y, Cheng J. Hydrolysis of lignocellulosic materials for ethanol production: A review. Bioresource Technology. 2002;**83**:1-11

[8] Bessou C, Ferchaud F, Gabrielle B, Mary B. Biofuels, greenhouse gases and climate change: A review. Agronomy for Sustainable Development. 2011;**31**:1-79. DOI: 10.1051/agro/2009039

[9] Wang MQ, Han J, Haq Z, Tyner WE, Wu M, Elgowainy A. Energy and greenhouse gas emission effects of corn and cellulosic ethanol with technology improvements and land use changes. Biomass and Bioenergy. 2011;**35**:1885-1896. DOI: 10.1016/j.biombioe.2011.01.028

[10] Kaylen M, Van Dyne DL, Choi YS, Blase M. Economic feasibility of producing ethanol from lignocellulosic feedstocks. Bioresource Technology. 2000;**72**:19-32. DOI: 10.1016/S0960-8524(99)00091-7

[11] Spatari S, Zhang Y, Maclean HL. Life cycle assessment of switchgrass- and corn automobiles. Environmental Science and Technology. 2005;**39**:9750-9758. DOI: 10.1021/es048293

[12] Bharathiraja B, Jayamuthunagai J, Praveenkumar R, Vinotharulraj J, Vinoshmuthukumar P, Saravanaraj A. Bioethanol production from lignocellulosic materials - An overview. Journal of Science and Technology. 2014;**1**:20-29. DOI: 10.1016/j.renene.2011.06.045

[13] Licht F. Renewable Fuels Association, Ethanol Industry Outlook 2008-2013 Reports [Internet]. 2013. Available from: www.ethanolrfa.org/pages/annual-industry-outlook

[14] Baker B. Global renewable fuel alliance [Internet]. 2014. Available from: http://globalrfa.org/news-media/global-ethanol-production-will-rise-to-over-90billion-litres-in

[15] Wang M, Han J, Dunn JB, Cai H, Elgowainy A. Well-to-wheels energy use and greenhouse gas emissions of ethanol from corn, sugarcane and cellulosic biomass for US use. Environmental Research Letters. 2012;**7**:1-13(045905). DOI: 10.1088/1748-9326/7/4/045905

[16] Demirbas A. Producing and using bioethanol as an automotive fuel. Energy Sources, Part B: Economics, Planning, and Policy. 2007;**2**:391-401. DOI: 10.1080/15567240600705466

[17] Farrell AE, Plevin RJ, Turner BT, Jones AD, O'Hare M, Kammen DM. Ethanol can contribute to energy and environmental goals. Science. 2006;**311**:506-508. DOI: 10.1126/science.1121416

[18] Schmer MR, Vogel KP, Mitchell RB, Perrin RK. Net energy of cellulosic ethanol from switchgrass. Proceedings of the National Academy of Sciences. 2008;**105**:464-469. DOI: 10.1073/pnas.0704767105

[19] Borrion AL, McManus MC, Hammond GP. Environmental life cycle assessment of lignocellulosic conversion to ethanol: A review. Renewable and Sustainable Energy Reviews. 2012;**16**:4638-4650. DOI: 10.1016/j.rser.2012.04.016

[20] MacLean HL, Spatari S. The contribution of enzymes and process chemicals to the life cycle of ethanol. Environmental Research Letters. 2009;**4**:1-10(014001). DOI: 10.1088/1748-9326/4/1/014001

[21] Monti A, Barbanti L, Zatta A, Zegada-Lizarazu W. The contribution of switchgrass in reducing GHG emissions. GCB Bioenergy. 2012;**4**:420-434. DOI: 10.1111/j.1757-1707.2011.01142.x

[22] Mu D, Seager T, Rao PS, Zhao F. Comparative life cycle assessment of lignocellulosic ethanol production: Biochemical versus thermochemical conversion. Environmental Management. 2010;**46**:565-578. DOI: 10.1007/s00267-010-9494-2

[23] Scown CD, Nazaroff WW, Mishra U, Strogen B, Lobscheid AB, Masanet E, Santero NJ, Horvath A, McKone TE. Erratum: Lifecycle greenhouse gas implications of US national scenarios for cellulosic ethanol production. Environmental Research Letters. 2012;**7**: 1-9(019502). DOI: 10.1088/1748-9326/7/1/019502

[24] Whitaker J, Ludley KE, Rowe R, Taylor G, Howard DC. Sources of variability in greenhouse gas and energy balances for biofuel production: A systematic review. GCB Bioenergy. 2010;**2**:99-112. DOI: 10.1111/j.1757-1707.2010.01047.x

[25] DiPardo J. Outlook for Biomass Ethanol Production and Demand. Washington: Energy Information Administration, Department of Energy; 2000

[26] Graf A, Koehler T. An evaluation of the potential for ethanol production in Oregon using cellulose-based feedstocks. Oregon Cellulose-Ethanol Study. 2000:1-30

[27] Flores-Cosío G, Arellano-Plaza M, Gschaedler-Mathis A, Amaya-Delgado L. Physiological response to furan derivatives stress by *Kluyveromyces marxianus* SLP1 in ethanol production. Revista mexicana de ingeniería química. 2018;**17**:189-202

[28] Ramadoss G, Muthukumar K. Influence of dual salt on the pre-treatment of sugarcane bagasse with hydrogen peroxide for bioethanol production. Chemical Engineering Journal. 2015;**260**:178-187. DOI: 10.1016/j.cej.2014.08.006

[29] Sandoval-Nuñez D, Arellano-Plaza M, Gschaedler AJ, Amaya-Delgado L. A comparative study of lignocellulosic ethanol productivities by *Kluyveromyces marxianus* and *Saccharomyces cerevisiae*. Clean Technologies and Environmental Policy. 2017:1-9. DOI: 10.1007/s10098-017-1470-6

[30] Mussato SI, Machado EMS, Carneiro LM, Teixeira JA. Sugars metabolism and ethanol production by different yeast strains from coffee industry wastes hydrolysates. Applied Energy. 2012;**92**:763-768. DOI: 10.1016/j.apenergy.2011.08.020

[31] Sathesh-Prabu C, Murugesan AG. Potential utilization of sorghum field waste for fuel ethanol production employing *Pachysolen tannophilus* and *Saccharomyces cerevisiae*. Bioresource Technology. 2011;**102**:2788-2792. DOI: 10.1016/j.biortech.2010.11.097

[32] Kim J, Jayoung R, Young H, Soon-Kwang H, Hyun AK, Yong KC. Ethanol production from galactose by a newly isolated *Saccharomyces cerevisiae* KL17. Bioprocess and Biosystems Engineering. 2014;**37**:1871-1878. DOI: 10.1007/s00449-014-1161-1

[33] Choi GW, Um HJ, Kang HW, Kim Y, Kim M, Kim YH. Bioethanol production by a flocculent hybrid, CHFY0321 obtained by protoplast fusion between *Saccharomyces cerevisiae* and *Saccharomyces bayanus*. Biomass and Bioenergy. 2010;**34**:1232-1242. DOI: 10.1016/j.biombioe.2010.03.018

[34] Kumari R, Pramanik K. Bioethanol production from Ipomoea Carnea biomass using a potential hybrid yeast strain. Applied Biochemistry and Biotechnology. 2013;**171**:771-785. DOI: 10.1007/s12010-013-0398-5

[35] Yang X, Xu M, Yang ST. Metabolic and process engineering of *Clostridium cellulovorans* for biofuel production from cellulose. Metabolic Engineering. 2015;**32**:39-48. DOI: 10.1016/j.ymben.2015.09.001

[36] Tomás-Pejó E, Oliva JM, González A, Ballesteros I, Ballesteros M. Bioethanol production from wheat straw by the thermotolerant yeast *Kluyveromyces marxianus* CECT 10875 in a simultaneous saccharification and fermentation fed-batch process. Fuel. 2009;**88**:2142-2147. DOI: 10.1016/j.fuel.2009.01.014

[37] Mehmood N, Alayoubi R, Husson E, Jacquard C, Büchs J, Sarazin C, Gosselin I. *Kluyveromyces marxianus*, an attractive yeast for ethanolic fermentation in the presence of imidazolium ionic liquids. International Journal of Molecular Sciences. 2018;**19**:887. DOI: 10.3390/ijms19030887

[38] Moreno AD, Ibarra D, Ballesteros I, González A, Ballesteros M. Comparing cell viability and ethanol fermentation of the thermotolerant yeast *Kluyveromyces marxianus* and *Saccharomyces cerevisiae* on steam-exploded biomass treated with laccase. Bioresource Technology. 2013;**135**:239-245. DOI: 10.1016/j.biortech.2012.11.095

[39] Pilap W, Thanonkeo S, Klanrit P, Thanonkeo P. The potential of the newly isolated thermotolerant *Kluyveromyces marxianus* for high-temperature ethanol production using sweet sorghum juice. Biotechnology. 2018;**8**:126. DOI: 10.1007/s13205-018-1161-y

[40] Mohd Azhar SH, Abdulla R, Jambo SA, Marbawi H, Gansau JA, Mohd Faik AA, Rodrigues KF. Yeasts in sustainable bioethanol production: A review. Biochemistry and Biophysics Reports. 2017;**10**:52-61. DOI: 10.1016/j.bbrep.2017.03.003

[41] Lara-Hidalgo C, Grajales-Lagunes A, Ruiz-Cabrera MA, Ventura-Canseco C, Gutiérrez-Miceli FA, Ruiz-Valdiviezo VM, Archila MA, Guillén-Navarro K, Sánchez JE. Agave Americana honey fermentation by *Kluyveromyces marxianus* "comiteco" production, a spirit from mexican southeast. Revista Mexicana de Ingeniería Química. 2017;**16**:771-779

[42] Nambu-Nishida Y, Nishida K, Hasunuma T, Kondo A. Development of a comprehensive set of tools for genome engineering in a cold- and thermo-tolerant *Kluyveromyces marxianus* yeast strain. Scientific Reports. 2017;**7**:1-7. DOI: 10.1038/s41598-017-08356-5

[43] Flores JA, Gschaedler A, Amaya-Delgado L, Herrera-López EJ, Arellano M, Arrizon J. Simultaneous saccharification and fermentation of agave tequilana fructans by *kluyveromyces marxianus* yeasts for bioethanol and tequila production. Bioresource Technology. 2013;**146**:267-273. DOI: 10.1016/j.biortech.2013.07.078

[44] Jung YH, Kim S, Yang J, Seo JH, Kim KH. Intracellular metabolite profiling of *Saccharomyces cerevisiae* evolved under furfural. Microbial Biotechnology. 2017;**10**:395-404. DOI: 10.1111/1751-7915.12465

[45] Piotrowski JS, Zhang Y, Bates DM, Keating DH, Sato TK, Ong IM, Landick R. Death by a thousand cuts: The challenges and diverse landscape of lignocellulosic hydrolysate inhibitors. Frontiers in Microbiology. 2014;**5**:1-8. DOI: 10.3389/fmicb.2014.00090

[46] Gao J, Feng H, Yuan W, Li Y, Hou S, Zhong S, Bai F. Enhanced fermentative performance under stresses of multiple lignocellulose-derived inhibitors by overexpression of a typical 2-Cys peroxiredoxin from *Kluyveromyces marxianus*. Biotechnology for Biofuels. 2017;**10**:1-13. DOI: 10.1186/s13068-017-0766-4

[47] Dörmo N, Bélafi-Bakó K, Bartha L, Ehrenstein U, Gubicza L. Manufacture of an environmental-safe biolubricant from fusel oil by enzymatic esterification in solvent-free system. Biochemical Engineering Journal. 2004;**21**:229-234. DOI: 10.1016/j.bej.2004.06.011

[48] Güvenç A, Kapucu N, Kapucu H, Aydoğan Ö, Mehmetoğlu Ü. Enzymatic esterification of isoamyl alcohol obtained from fusel oil: Optimization by response surface methodolgy. Enzyme and Microbial Technology. 2007;**40**:778-785. DOI: 10.1016/j.enzmictec.2006.06.010

[49] Ozyilmaz G, Yagiz E. Isoamylacetate production by entrapped and covalently bound *Candida rugosa* and porcine pancreatic lipases. Food Chemistry. 2012;**135**:2326-2332. DOI: 10.1016/j.foodchem.2012.07.062

[50] Löser C, Urit T, Bley T. Perspectives for the biotechnological production of ethyl acetate by yeasts. Applied Microbiology and Biotechnology. 2014;**98**:5397-5415. DOI: 10.1007/s00253-014-5765-9

[51] Ehrlich F. Über das natürliche Isomere des Leucins. Berichte der deutschen chemischen Gesellschaft. 1907;**40**:1027-1047

[52] Hazelwood LA, Daran JM, Van Maris AJA, Pronk JT, Dickinson JR. The Ehrlich pathway for fusel alcohol production: A century of research on *Saccharomyces cerevisiae* metabolism. Applied and Environmental Microbiology. 2008;**74**(8):2259-2266. DOI: 10.1128/AEM.00934-08

[53] Morrissey JP, Etschmann MM, Schrader J, Billerbeck G. Cell factory applications of the yeast *Kluyveromyces marxianus* for the biotechnological production of natural flavour and fragrance molecules. Yeast. 2015;**32**:3-16. DOI: 10.1002/yea.3054

[54] Amaya-Delgado L, Herrera-López EJ, Arrizon J, Arellano-Plaza M, Gschaedler A. Performance evaluation of *Pichia kluyveri, Kluyveromyces marxianus* and *Saccharomyces cerevisiae* in industrial tequila fermentation. World Journal of Microbiology and Biotechnology. 2013;**29**:875-881. DOI: 10.1007/s11274-012-1242-8

Chapter 2

Second-Generation Bioethanol Production through a Simultaneous Saccharification-Fermentation Process Using *Kluyveromyces Marxianus* Thermotolerant Yeast

Jorge A. Mejía-Barajas, Mariana Alvarez-Navarrete,
Alfredo Saavedra-Molina, Jesús Campos-García,
Uri Valenzuela-Vázquez,
Lorena Amaya-Delgado and Melchor Arellano-Plaza

Additional information is available at the end of the chapter

http://dx.doi.org/10.5772/intechopen.78052

Abstract

Due to the present renewable fuels demand increase, reduction of second-generation bioethanol production cost is pursued, since it is considered the most promising biofuel, but not yet economically viable. A proposed solution is its production through a simultaneous saccharification and fermentation process (SSF); however, it is necessary to apply temperatures above 40°C, which reduce the viability of traditional ethanologenic yeasts. As consequence, the use of thermotolerant ethanologenic yeast has been suggested, among which the yeast *Kluyveromyces marxianus* stands out. This chapter addresses the production of second-generation bioethanol through the SSF process, emphasizing the potential of *K. marxianus* to transform lignocellulosic biomass as agave bagasse. As result, it is proposed to direct the second-generation bioethanol production to the SSF process employing thermotolerant yeasts, to increase process productivity, and addressing the economic barriers.

Keywords: bioethanol, simultaneous saccharification and fermentation (SSF), thermotolerant yeasts, Kluyveromyces marxianus, agave bagasse

1. Introduction

The consumption of fossil fuels derived from petroleum is one of the main sources of pollution of the environment, in addition to its expensive and decreasing production, whereas its demand is increasing [1]. This is why countries around the world have directed their policies

toward the biofuels usage, which are sustainable, biodegradable, with high combustion efficiency, and their development generates manufacturing and investment jobs, promoting the de agricultural sector development, as well reducing greenhouse gases [2, 3]. This way, the use of biofuels such as bioethanol is pursued to reduce dependence on fossil fuels and contribute to meet the future demands of energy in the world, and at the same time meeting the carbon dioxide emissions reduction goals specified in the Kyoto Protocol [4]. Therefore, it is expected that by 2050 biofuels contribute 30% of the world's fuel demand [5].

Economically viable bioethanol production still has to date challenges to overcome. This chapter addresses the lignocellulosic biomass utilization for second-generation bioethanol production through a simultaneous saccharification and fermentation process, utilizing thermotolerant yeasts such as *K. marxianus*.

2. Bioethanol

Bioethanol is one of the most used biofuels with a worldwide production of around 27 billion gallons per year [2, 6]. This biofuel, defined as ethanol produced from biomass has characteristics such as low combustion temperature, high octane number, and lower evaporation loss compared to gasoline [7, 8]. Disadvantages of bioethanol compared to gasoline are its lower energy density and vapor pressure, as well as water miscibility and corrosive capacity [9].

Bioethanol can be mixed with gasoline in 10% (E10), 20% (E20), and 22% (E22) proportions, without the need to make mechanical modifications in combustion vehicles [9]. There are even current designs by some manufacturers that allow vehicles to use up to 85% ethanol [10] and in Brazil more than 20% of cars can use 100% ethanol as fuel [2]. The main purpose of bioethanol, when mixed with gasoline is as an oxygenating agent. Mixed with gasoline, ethanol provides advantages such as increased gas volume change, better combustion, and reduced carbon dioxide emission [11]. It has also been shown that bioethanol can significantly reduce SO_2 emissions when mixed with 95% gasoline. This is because the fuel added with bioethanol increases its oxygen content, causing a better oxidation of hydrocarbons and decreasing the emission of greenhouse gases [4].

The main bioethanol producing countries are currently the United States and Brazil, generating up to 70% of world production [12]. However, the bioethanol industry has expanded to other countries such as China, Argentina, and the European Union due to this product increased demand [13]. In the case of the United States, there has been a dramatic increase in bioethanol production from 175 million gallons in 1980 to 14,810 million gallons in 2015 [14].

2.1. First- and second-generation bioethanol

Bioethanol is currently obtained in commercial quantities mainly from the fermentation of simple sugars using food inputs such as corn, sugar cane, and sorghum as raw material. The bioethanol obtained from this class of substrates is called first generation bioethanol [15]. The viability of the production of first-generation biofuels is questionable, due to their associated conflicts, such as ethical aspects and their high-cost since the raw materials are linked to the food market, which affects the final price of the product [16].

Given the problems of first-generation fuels, an alternative would be second-generation fuels, where fermentable sugars are derived from lignocellulosic biomass, which are present in agro-industrial wastes. By using industrial wastes as a raw material, pollution is reduced by the elimination of these potentially polluting wastes, as well the materials being of low-cost and its handling and conservation is efficient and economical [17]. Biofuel production such as second-generation bioethanol is considered one of the most promising strategies to replace non-renewable fossil fuels because it does not interfere with the materials available for human or animal consumption, at the same time as collaborating with sustainable development [18, 19].

As a disadvantage, a technological investment is necessary for the treatment of lignocellulosic biomass, and currently, its production is not economically sustainable [20, 21].

Currently, second-generation bioethanol is produced mainly in pilot plants and most commercial plants have been built in the last decade in Denmark, Finland, Spain and Italy, and the United States [22] and due to the challenges that its commercialization still represents the design and optimization of different processes for second-generation bioethanol production has been promoted to reduce production costs.

3. Process configuration for second-generation bioethanol production

There are different process configurations for bioethanol production, but all of them include the steps of raw material pretreatment to achieve biomass components solubilization and separation (cellulose, hemicellulose, and lignin); lignocellulosic material hydrolysis to degrade its components and obtain simple sugars; and the fermentation of the substrate to transform the sugars into bioethanol. Reported processes vary mainly in the number of stages and bioreactors needed, which present different pH conditions, oxygenation, sugar concentration, and temperature. **Figure 1** shows the main process configurations and their stages, while **Table 1** lists the main characteristics of each one of these configurations [23].

The CBP proposes cellulase enzyme production by microorganisms integrated into the fermentation, reducing the enzyme cost in the bioethanol production. However, currently, this process is in its early stages of development since a limited number of microorganisms capable of generating economically viable enzymes are reported. Furthermore, there is no microorganism or microorganism consortium that generates cost-effective bioethanol through CBP at the industry level [29]. In consideration to this situation, the development of genetically modified microorganisms able to produce these enzymes with an economically viable concentration and activity is one of the most promising options. Within the genetically modified microorganisms, yeasts have been one of the most used. Hasunuma and Kondo [30] presented a review of the development of yeast cells for second-generation bioethanol production through CBP. Within their study, they conclude that the combination of cell surface

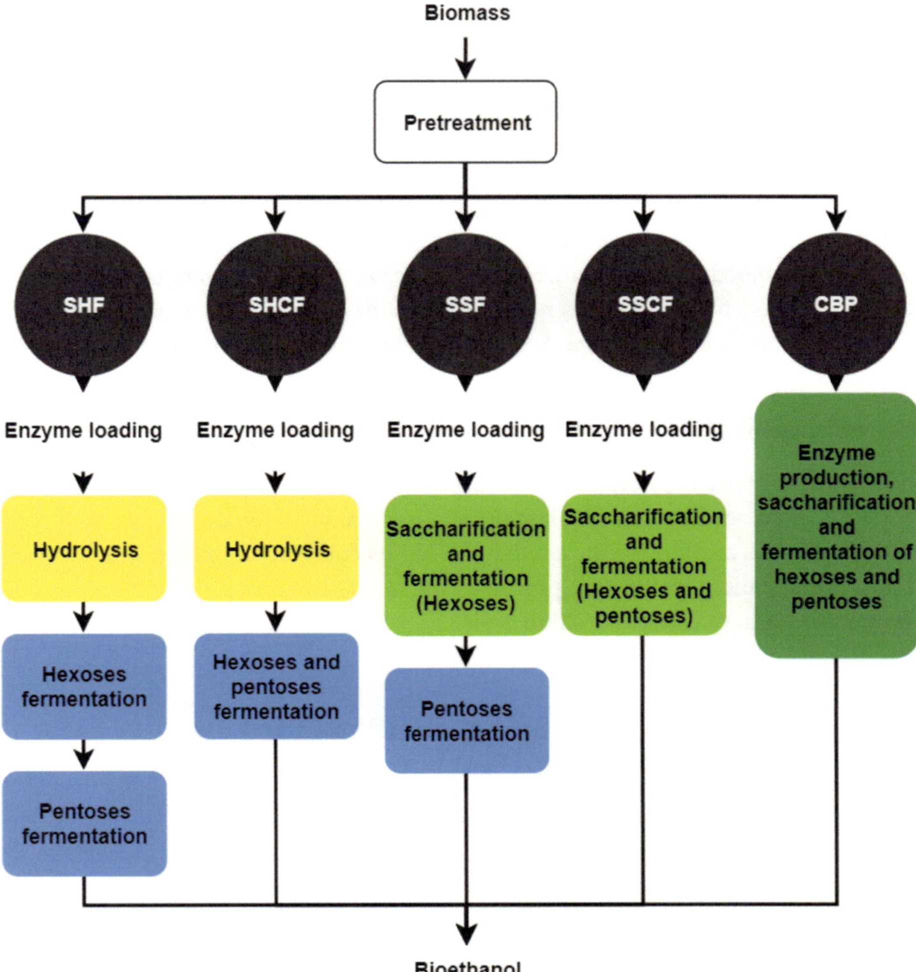

Figure 1. Processes for second-generation bioethanol production. Each box refers to a bioreactor. The pretreatment consists of a chemical hydrolysis. Sequential hydrolysis and fermentation (SHF); sequential hydrolysis and cofermentation (SHCF); simultaneous saccharification-fermentation (SSF); SSCF: Simultaneous saccharification-cofermentation (SSCF); consolidated bioprocess (CBP).

engineering and metabolome are an efficient proposal for the development of CBP yeasts strains. Favaro et al., Mattam et al., Liao et al., and Van et al. [31–34] review the possibilities of recombinant yeasts generation for second-generation bioethanol production through CBP.

In the rest of the mentioned processes in **Figure 1**, the cellulases are added beforehand SHF or during the fermentation process SSF. These enzymes can be purchased commercially through different companies responsible for selecting enzymes with the best characteristics to perform these processes. The current commercial cellulases are produced mainly by fungi, bacteria, and yeasts, although they can also be produced by plants and ruminants [35].

Process	Acronym	Main characteristics	Strain, substrate, temperature, % yield
Sequential hydrolysis and fermentation	SHF	The biomass hydrolysis and sugar fermentation are carried out sequentially in separate bioreactors. In this process the optimum temperatures for enzymatic hydrolysis and fermentation are different and it is necessary to cool the hydrolysate after the enzymatic hydrolysis to start fermentation. To achieve pentoses consumption it is necessary to add microorganisms capable of converting pentoses into ethanol at the end of hexoses fermentation. This is one of the most expensive processes, with long times, where enzymatic inhibition by the sugars generated can be present.	*Saccharomyces cerevisiae* 11.25% Bambú 30°C, 41% Sindhu *et al.*[24]
Sequential hydrolysis and cofermentation	SHCF	Unlike the SHF process, in this process the hydrolysis and fermentation of hexoses and pentoses is carried out in the same bioreactor. It is necessary the use of yeasts with the capacity to metabolize pentoses or the addition of a microbial consortium that together manage to assimilate both hexoses and pentoses in the medium.	*Scheffersomyces shehatae* 10% Bagazo caña 30°C, 82% Dussan *et al.* [25]
Simultaneous saccharification-fermentation	SSF	Combines the enzymatic hydrolysis and fermentation of hexoses in a single bioreactor simultaneously, increasing the efficiency of enzymatic hydrolysis by eliminating product inhibition, decreasing the risk of microbial contamination by being a single process and increasing bioethanol productivity compared with the SHF process. The pH and temperature values for these processes have been shown to be very rigid since the optimum pH and temperature values of the enzymes and yeasts used are compromised. At the end of the fermentation of hexoses, the addition of microorganisms capable of converting pentoses into ethanol is required.	*Kluyveromyces marxianus* 10% Pulpa de Zanahoria 42°C, 92% Yu *et al.*[26]
Simultaneous saccharification-cofermentation	SSCF	This process is similar to SSF with the difference that it uses yeasts capable of fermenting both hexoses and pentoses in the same bioreactor. This promotes the use of a greater amount of substrate, which can increase the production of ethanol and productivity increase by reducing the times of each stage.	*Saccharomyces cerevisiae* 10% Pícea 34°C, 85% Bertilsson *et al.*[27]
Consolidated bioprocess	CBP	The enzyme production, enzymatic hydrolysis and fermentation of the sugars are integrated into the same bioreactor. As result, the addition of external cellulases is not necessary.	*Scheffersomyces shehatae* 10% Almidón 30°C, 29% Tanimura *et al.*[28]

Table 1. Main characteristics of the second-generation bioethanol production processes accompanied by an example of its implementation in the literature [23–28].

3.1. Sequential hydrolysis and fermentation (SHF)

This process is carried out in two stages, where hydrolysis and fermentation operate in different procedures. First, the enzymatic cocktail is used to hydrolyze the pretreated lignocellulosic biomass to obtain sugar monomers. The resulting hydrolyzate is subsequently used as a substrate for the fermentation process of sugars to ethanol [36]. The cellulose hydrolysis process through cellulases is the most feasible method for the liberation of sugars since in optimum conditions yields greater than 90% can be obtained. Chandel et al. [37] reviewed the techniques developed in molecular biology and cellulase engineering, as well as the application of cellulases for cellulose hydrolysis.

The main disadvantage of the SHF process is that both stages operate in their respective optimal conditions, thus more processing time is necessary. In addition, the hydrolytic enzymes employed can suffer from product inhibition. These characteristics impact on the productivity of the process [25]. The enzyme cost contributes significantly to the second-generation bioethanol final price, which is why more research is needed in order to reduce saccharification costs through the use of cellulases [36, 37]. In order to make second-generation bioethanol production affordable, cellulase cost must be decreased, and one solution is to increase its activity, which can be achieved through SSF processes at optimum temperatures of the enzymes employed [38].

3.2. Simultaneous saccharification-fermentation (SSF)

The main characteristic of the SSF process is that the stages of enzymatic hydrolysis and fermentation are carried out simultaneously. This reduces the energy investment, and therefore, the operating costs, in addition to optimizing the process by reducing the time needed for each of them. These stages favor a greater enzymatic activity of the cellulases eliminating product inhibition since the sugars are metabolized by the yeasts simultaneously as they are released in the hydrolysis. In general, through the SSF process, higher ethanol yields have been obtained compared to SHF, with increases from 13 to 30% [39]. The main disadvantage of the SSF process is the cellulase activity optimal temperature (45–60°C), which is higher than that required for the yeast growth and fermentation (30–35°C) [40, 41]. Besides, fermentation is an exothermic process, so as the fermentation progresses the temperature increases [42]. Therefore, the temperature is one of the main factors that must be considered when establishing a SSF system.

3.3. Effect of temperature in SSF process

Fermentation at high temperatures presents advantages such as bioreactor cooling costs reduction and ethanol extraction promotion, reducing its toxic effects on yeasts. Moreover, it is possible to do this process in warm climate countries [43]. In a study conducted by Abdel-Banat et al., an increase of 5°C in the fermentation process, a reduction of enzyme cost up to 50% was observed. However, high temperatures generate yeast growth inhibition, decrease in the cell cycle, increase in fluidity and reduction of the plasma membrane permeability, intracellular pH reduction, breakage of cytoskeleton filaments and microtubules, proteins synthesis repression, mutation frequency increment, and inefficient damaged DNA repair. All the above effects reduce the yeast viability and decrease the bioethanol production yield. Therefore, the use of thermotolerant yeasts in the SSF process for second-generation bioethanol production is proposed as a promising option [23].

4. Yeasts in SSF processes

Although traditionally *S. cerevisiae* yeasts have been the most used in fermentation processes, the production of second-generation bioethanol confronts these microorganisms to conditions not found in traditional fermentation processes [44, 45].

In a SSF process, the selected yeast must be a thermotolerant strain. Thermotolerant yeasts are those that have an optimal growth at temperatures equal to or greater than 40°C [46]. During the last years, potential industrial applications of thermotolerant yeasts have been developing, such as prebiotic and probiotic agents, biomass, and recombinant protein production, as well as bioethanol production [47]. Bioethanol production through SSF process using thermotolerant yeasts generates a reduction in investment costs, such as the industrial equipment needed, lower contamination degree, and decreased process time [17].

The most known non-*Saccharomyces* yeast species used in SSF processes are *K. marxianus*, although there are also reports of other species such as *K. fragilis*, *H. polymorpha*, and *P. pastoris* [47].

4.1. Kluyveromyces marxianus

K. marxianus strains are phenotypically very diverse due to the great variety of habitats in which they have been isolated, resulting in a great metabolic diversity [48].

In general, they are considered GRAS (Generally recognized as safe), they are the eukaryotic cells that have presented the highest growth rate [49], they have an efficient ethanol production capacity up to 45°C, with thermotolerance up to 52°C, besides the genomes of some strains have been described [50–56]. Recently, studies have been carried out on the optimization of the metabolic engineering pathways in these yeasts [57], and genetic engineering has been used to obtain strains capable of producing heterologous proteins or metabolites such as lactate and xylitol [58, 59].

One necessary characteristic in sustainable bioethanol production is the fermentation of different sugars [60], innate in most of *K. marxianus* yeasts. These yeasts can ferment xylose, xylitol, cellobiose, lactose, and arabinose, both in liquid and solid medium, considered a great advantage compared with *S. cerevisiae* [61].

Whereas strains of *S. cerevisiae* have been obtained by genetic engineering with pentose metabolism [62], these strains still present different problems that must be solved [63], besides that they are not thermotolerant. Nitiyon et al. [64] reported that the yeast *K. marxianus* BUNL-21 presents a xylose to ethanol efficient conversion capacity, as well as thermotolerance. López-Alvarez et al. [65] obtained higher ethanol yields with *K. marxianus* UMPe-1 yeast compared with *S. cerevisiae* Pan-1. Lyubomirov et al. [66] and Kuloyo et al. [67] compared the ethanol production at temperatures of 35 and 40°C by the strains *K. marxianus* UOFS Y-2791 and *S. cerevisiae* UOFS Y-0528, concluding *K. marxianus* presents potential as an alternative to *S. cerevisiae* for the bioethanol production, as well as other metabolites such as 2-phenyl ethanol.

Due to the aforementioned characteristics, *K. marxianus* have been considered as one of the yeast with the highest potential for the second-generation bioethanol production, and a viable alternative compared with *S. cerevisiae* [68, 30].

4.2. *K. marxianus* in the bioethanol production through SSF

As a thermotolerant yeast and due to its ability to use various sugars as a carbon source, *K. marxianus* yeasts have been used widely for second-generation bioethanol production through SSF and SSCF processes (**Table 2**).

Kádár et al. [80] compared the yield in the second-generation bioethanol production by *K. marxianus* and *S. cerevisiae* yeasts, in an SSF process at 40°C. Having found no significant differences in the ethanol production with respect to SHF processes, it is suggested to carry out the fermentation processes with *K. marxianus* thermotolerant yeasts at temperatures above 40°C. Tomás-Pejó et al. [79] performed SSF processes with a *K. marxianus* thermotolerant yeast CECT 10,875 at 50°C. By using a feed back process they increased ethanol production by 20%. Hyun-Woo et al. [75] carried out an SSF process with a temperature change from 45 to 35°C at 24 h of the process using the thermotolerant yeast *K. marxianus* CHY 1612. This change generated an increase of 12 g/L of ethanol, compared to a SSF process carried out at a constant temperature of 45°C. Yu-Sheng et al. [73] studied the bioethanol production using a *K. marxianus*

Strain	Temperature (°C)	Carbon source	Ethanol (g/L)	g EtOH/g substrate	Productivity (g/L/h)	Reference
K 213	42	Water hyacinth	7	0.14	0.31	Yan *et al.*, 2015
MTCC 1338	42	*Moringa oleifera* softwood	16-21	-	0.19	Shanmugam *et al.*, 2014
NRRL Y-6860	45	Rice straw	11	0.24	1.44	Cunha *et al.*, 2014
UFV-3, ATCC 8554 y CCT 4086	37,42	Sugar cane bagasse	13-22	0.17-0.29	1.62-2.75	Costa *et al.*, 2014
BCRC 21363	37-45	Sugar cane bagasse	22-24	0.22-0.24	0.30-0.33	Yu-Sheng *et al.*, 2013
K 21	42	Carrot pomace	18-37	0.18	0.75-0.88	Chi-Yang *et al.*, 2013
CECT 10875	42	Wheat straw	11	0.18	0.36	Moreno *et al.*, 2013
CHY 1612	35,45	Barley straw	22-34	0.13-0.21	0.30-0.47	Hyun-Woo *et al.*, 2011
IMB 3	45	*Kanlow* switchgrass	22-32	0.27	0.30-0.44	Pessani et *al.*, 2011
CECT 10875	42	Barley straw	19-29	0.19	0.66-0.77	García-Aparicio *et al.*, 2011
IMB 1, IMB 2, IMB 3, IMB 4, y IMB 5	45	Kanlow Switchgrass	16-21	0.13-0.17	0.22-0.29	Faga *et al.*, 2010
6556, 397 y 2762	37	Corncob, soybean cake	2-5	0.04	0.10	Zhang *et al.*, 2010
CECT 10875	50	Wheat straw	22-36	0.22-0.25	0.50-1.18	Tomás-Pejó *et al.*, 2009
IMB 4	37-45	Kanlow switchgrass	12-16	0.16-.21	0.17-0.22	Suryawati *et al.*, 2008
Y 01070	40	Paper sludge	8-17	0.13-0.28	0.11-0.24	Kádár y Réczey, 2004
CECT 10875	42	*Eucalyptus globulus*, Sorghum and wheat straw	16-19	0.16-0.19	0.22-0.26	Ballesteros *et al.*, 2004
ITS4SC	42	Rice and wheat Straw, Sugar cane bagasse	18-23	0.18-0.23	0.30-0.38	Madhurinarra *et al.* 2015
CK8	43	Rice husk	15	0.16	0-16	Nachaiwieng *et al.* 2015
K21	40	Taro waste	49	0.29	2.23	Wu *et al.* 2016

Table 2. Second-generation bioethanol production using ***K. marxianus*** thermotolerant yeasts in SSF processes [26, 50, 68–84].

thermotolerant yeast, through the SSF process in a rotating reactor, which allowed a constant exchange of biomass that was in contact with the yeast, concluding that through this process bioethanol production has commercial potential. Wu et al. [84] implemented a SSF process with a high solid load of taro waste using *K. marxianus*, reaching 94% of theoretical yields in 20 h of fermentation, which was reflected in high process productivity (Wu et al. [84]).

With the previous reports, we observed that modifications to the SSF process using *K. marxianus* thermotolerant yeasts can increase ethanol production to economically viable levels.

5. Substrate selection for second-generation bioethanol production through SSF: agave bagasse case

Lignocellulosic biomass is a source of renewable energy, available in most of the world. However, its treatment is one of the main factors that increase the cost of second-generation bioethanol production. The biomass selection for this process is directly correlated with its availability in the production area, characteristics that depend on geographical variables [85]. In Mexico, agave bagasse is one of the most generated lignocellulosic materials, since it is an agro-industrial waste resulting from tequila and mezcal production. The blue agave (*Agave tequilana*) used for tequila production is cultivated mainly in the western region of Mexico. In general, the process of tequila production considers the use of blue agave plant cores, which are cooked in ovens or systems such as the diffuser. Afterward, they are pressed for their juice extraction, and the fructans present in the juice are then hydrolyzed to monosaccharides. The residue of this process is agave bagasse. It is estimated that 859,000 tons of agave are processed per year to produce tequila, and approximately 343,600 tons of agave bagasse are generated. Agave bagasse can be used as livestock food, construction material, and for recycled paper elaboration [86], as well as a substrate for edible fungi growth [87]. However, most of it is incinerated, which generates large amounts of ash that can contaminate rivers, bodies of water and damage flora and fauna [88]. **Table 3** shows that agave bagasse has a higher cellulose proportion, compared to main lignocellulosic biomass used for bioethanol production.

Hernández-Salas et al. [90] obtained a sugar yield of 12–58% by hydrolysis of agave bagasse using an alkaline-enzymatic treatment, while under the same conditions with sugarcane bagasse the yield was lower, with values of 11–20% [90]. Therefore, according to its production and composition, agave bagasse can be considered a promising source of fermentable sugars for bioethanol production. **Table 4** shows studies for bioethanol production using agave bagasse.

Caspeta et al. [92] released 91% of agave bagasse sugars during saccharification and produced 64 g/L of ethanol after 9 h of fermentation with *S. cerevisiae* SuperStart yeast, this being the highest yield obtained with agave bagasse.

Lignocellulosic biomass	Cellulose (%)	Hemicellulose (%)	Lignin (%)
Agave bagasse	42.0	20.0	15.0
Sugar cane bagasse	40.0	27.0	10.0
Corn stover	35.0	14.4	21.5
Corncob	33.7	31.9	6.1
Wheat straw	32.9	24.0	8.9

Table 3. The lignocellulosic composition of agroindustrial wastes used in second-generation bioethanol production [89].

Yeast	Process	Ethanol (g/L)	Yield	Productivity[a]	Reference
S. cerevisiae	SHF	7	0.12	0.12	Hernández-Salas *et al.*, 2009
Pichia caribbica UM-5	SHF	18	0.18	0.15	Saucedo-Luna *et al.*, 2011
S. cerevisiae SuperStart	SHF	64	0.25	1.33	Caspeta *et al.*, 2014
S. cerevisiae Thermosacc	SHF	31	0.15	0.52	Montiel *et al.*, 2016
S. cerevisiae ATCC 4126	SHF	65	0.26	0.77	Rios González *et al.* 2017

Productivity is obtained considering total time of hydrolysis and fermentation.

Table 4. Bioethanol production from agave bagasse [90–94].

Rios González et al. [94] managed to implement a process of autohydrolysis pretreatment which allowed preserving the glycan content in agave bagasse, achieving a high digestibility in the hydrolysis process for its subsequent fermentation to ethanol with a strain of *S. cerevisiae*.

Through a simulation program analysis, Barrera et al. [95] carried out the technical and economic evaluation of bioethanol production, considering sugarcane bagasse, and agave bagasse as lignocellulosic biomass substrates. The results showed a lower production cost using agave bagasse (1.34 USD/gallon), compared to sugarcane bagasse (1.46 USD/gallon), suggesting that this result is due to the lower processing cost required for agave bagasse and its low lignin content [95].

Agave bagasse, besides being a good source of sugars for bioethanol production, is considered one of the best agro-industrial residues generated in the Mexico region, to be used in solid state fermentation processes [3], as well as for succinic acid production [96].

It is worth highlighting the scarce reports of bioethanol production from agave bagasse using non-*Saccharomyces* strains, as well as there are only reported SHF processes with this material, which represents a study opportunity to use this substrate in more efficient processes such as SSF.

6. Conclusion

Dependence on fossil fuels has led to a high degree of pollution on the planet, as well as low availability and an increase in its price, which forces the pursuit of new sources of energy. The use of second-generation bioethanol is a promising option to face this problem. However, currently, its production is not affordable, which has prevented its commercialization. Although metabolic engineering in conjunction with bioprocess optimization is recommended techniques for bioethanol cost-effective production, these are still in development, which contrasts with the widely used and perfected yeast selection techniques. These approaches can be used to find thermotolerant yeasts such as *K. marxianus* for their application in the second-generation bioethanol production through SSF processes to overcome the economic challenges in the production of this biofuel.

Author details

Jorge A. Mejía-Barajas[1], Mariana Alvarez-Navarrete[2], Alfredo Saavedra-Molina[1], Jesús Campos-García[1], Uri Valenzuela-Vázquez[3], Lorena Amaya-Delgado[3] and Melchor Arellano-Plaza[3*]

*Address all correspondence to: marellano@ciatej.mx

1 Instituto de Investigaciones Químico-Biológicas, Universidad Michoacana de San Nicolás de Hidalgo, Morelia, México

2 Departamento de Ingeniería Química y Bioquímica, Instituto Tecnológico de Morelia, Morelia, México

3 Centro de Investigación y Asistencia en Tecnología y Diseño del Estado de Jalisco, A.C. Guadalajara, México

References

[1] Jönsson L, Alriksson B, Nilvebrant N. Bioconversion of lignocellulose: Inhibitors and detoxification. Biotechnology for Biofuels. 2013;**6**:16-26

[2] Baeyens J, Kang Q, Appels L, Dewil R, Lv Y, Tan T. Challenges and opportunities in improving the production of bio-ethanol. Progress in Energy and Combustion Science. 2015;**47**:60-88

[3] Ramos JL, Valdivia M, García-Lorente F, Segura A. Benefits and perspectives on the use of biofuels. Microbial Biotechnology. 2016;**9**(4):436-440

[4] Gupta A, Verma JP. Sustainable bio-ethanol production from agro-residues: A review. Renewable and Sustainable Energy Reviews. 2015;**41**:550-567

[5] Guo M, Song W, Buhain J. Bioenergy and biofuels: History, status and perspective. Renewable & Sustainable Energy Reviews. 2015;**42**:712-725

[6] RFA, Ethanol industry outlook. Report by the Renewables Fuel Association. 2015. Available from: http://www.ethanolrfa.org/pages/statistics. [Accessed April 1, 2018]

[7] Ashfaq A, Kaifiyan M, Mohammad S. Environmental aspects of new generation fuels through energy balance, biodiversity and agriculture. International Journal of Current Research and Academic Review. 2015;**3**(4):15-19

[8] Ávila-Lara AI, Camberos-Flores JN, Mendoza-Pérez JA, Messina-Fernández SR, Saldaña-Duran CE, Jimenez-Ruiz EJ, Sánchez-Herrera LM, Pérez-Pimienta JA. Optimization of alkaline and dilute acid pretreatment of agave bagasse by response surface methodology. Frontiers in Bioengineering and Biotechnology. 2015;**3**:146-155

[9] Balatat M, Balata H, Cahide O. Progress in bioethanol processing. Progress in Energy and Combustion Science. 2008;**34**:551-573

[10] Haghighi Mood S, Hossein Golfeshan A, Tabatabaei M, Salehi Jouzani G, Najafi GH, Gholami M, Ardjmand M. Lignocellulosic biomass to bioethanol, a comprehensive review with a focus on pretreatment. Renewable and Sustainable Energy Reviews. 2013;**27**:77-93

[11] Ragauskas AJ, Williams CK, Davison BH, Britovsek G, Cairney J, Eckert CA, Frederick JWJ, Hallett JP, Leak DJ, Liotta CL. The path forward for biofuels and biomaterials. Science. 2006;**311**:484-489

[12] Argueso JL, Carazzolle MF, Mieczkowski PA, Duarte FM, Netto OV, Missawa SK, Galzerani F, Costa GG, Vidal RO, Noronha MF, Dominska M, Andrietta MG, Andrietta SR, Cunha AF, Gomes LH, Tavares FC, Alcarde AR, Dietrich FS, McCusker JH, Petes TD, Pereira GA. Genome structure of a Saccharomyces cerevisiae strain widely used in bioethanol production. Genome Research. 2009;**19**:2258-2270

[13] Awasthi P, Shrivastava S, Kharkwal AC, Varma A. Biofuel from agricultural waste: A review. International Journal of Current Microbiology and Applied Sciences. 2015;**4**(1):470-477

[14] Zabed H, Sahu JN, Suely A, Boyce AN, Faruq G. Bioethanol production from renewable sources: Current perspectives and technological progress. Renewable and Sustainable Energy Reviews. 2017;**71**:475-501

[15] Zabaniotou A, Ioannidou O, Skoulou V. Rapeseed residues utilization for energy and 2nd generation biofuels. Fuel. 2008;**87**:1492-1502

[16] Ullah K, Kumar Sharma V, Dhingra S, Braccio G, Ahmad M, Sofia S. Assessing the lignocellulosic biomass resources potential in developing countries: A critical review. Renewable and Sustainable Energy Reviews. 2015;**51**:682-698

[17] Ishola MM, y Taherzadeh MJ. Effect of fungal and phosphoric acid pretreatment on ethanol production from oil palm empty fruit bunches (OPEFB). Bioresource Technology. 2014;**165**:9-12

[18] Maitan-Alfenas GP, Visser EM, Guimarães VM. Enzymatic hydrolysis of lignocellulosic biomass: Converting food waste in valuable products. Current Opinion in Food Science. 2015;**1**:44-49

[19] Aimaretti N, Ybalo C, Escorcia M, Codevilla A. Revalorización De Descartes Agroindus triales Para La Obtención De Bioetanol. Invenio. 2012;**15**(28):141-157

[20] Tesfaw A, y Assefa F. Current trends in bioethanol production by Saccharomyces cerevisiae: Substrate, inhibitor reduction, growth variables, Coculture, and immobilization. International Scholarly Research Notices. 2014;**2014**:1-11

[21] Zhao L, Zhang X, Xu J, Ou X, Chang S, Wu M. Techno-economic analysis of bioethanol production from lignocellulosic biomass in China: Dilute-acid Pretreatment and enzymatic hydrolysis of corn Stover. Energies. 2015;**8**:4096-4117

[22] Paulova L, Patakova P, Branska B, Rychtera M, Melzoch K. Lignocellulosic ethanol: Technology design and its impact on process efficiency. Biotechnology Advances. 2015;**33**:1091-1107

[23] Choudhary J, Singh S, Nain L. Thermotolerant fermenting yeasts for simultaneous saccharification and fermentation of lignocellulosic biomass. Electronic Journal of Biotechnology. 2016;**21**:82-92

[24] Sindhu R, Binod P, Pandey A. Biological pretreatment of lignocellulosic biomass—An overview. Bioresource Technology. 2016;**199**:76-82

[25] Dussán KJ, Silva DDV, Perez VH, da Silva SS. Evaluation of oxygen availability on ethanol production from sugarcane bagasse hydrolysate in a batch bioreactor using two strains of xylose-fermenting yeast. Renewable Energy. 2016;**87**:703-710

[26] Yu CY, Jiang BH, Duan KJ. Production of bioethanol from carrot pomace using the thermotolerant yeast kluyveromyces marxianus. Energies. 2013;**6**(3):1794-1801

[27] Magnus B, Kim O, Gunnar L. Prefermentation improves xylose utilization in simultaneous saccharification and co-fermentation of pretreated spruce. Biotechnology for Biofuels. 2009;**2**(1):1-10

[28] Tanimura A, Kikukawa M, Yamaguchi S, Kishino S, Ogawa J, Shima J. Direct ethanol production from starch using a natural isolate, Scheffersomyces shehatae: Toward consolidated bioprocessing. Scientific Reports. 2015;**5**(9593):1-7

[29] Jouzani S, y Taherzadeh MJ. Advances in consolidated bioprocessing systems for bioethanol and butanol production from biomass: A comprehensive review. Biofuel Research Journal. 2015;**5**:152-195

[30] Hasunuma T. y Kondo A. Development of yeast cell factories for consolidated bioprocessing of lignocellulose to bioethanol through cell surface engineering. Biotechnology Advances 2012;**30**(6):1207-1218

[31] Favaro L, Jooste T, Basaglia M. Designing industrial yeasts for the consolidated bioprocessing of starchy biomass to ethanol. Bioengineered. 2013;**4**:97-102

[32] Mattam AJ, Kuila A, Suralikerimath N, Choudary N, Rao PVC, Velankar HR. Cellulolytic enzyme expression and simultaneous conversion of lignocellulosic sugars into ethanol and xylitol by a new Candida tropicalis strain. Biotechnology for Biofuels 2016;**9**:157-169

[33] Liao B, Hill GA, Roesler WJ. Stable expression of barley a-amylase in S. Cerevisiae for conversion of starch into bioethanol. Biochemical Engineering Journal. 2012;**64**:8-6

[34] Van WH, Bloom M, Viktor MJ. Engineering yeasts for raw starch conversion. Applied Microbiology and Biotechnology. 2012;**95**:1377-1388

[35] Park YW, Tominaga R, Sugiyama J. Enhancement of growth by expression of poplar cellulase in Arabidopsis thaliana. The Plant Journal. 2003;**33**:1099-1106

[36] Rastogi M, Shrivastava S. Recent advances in second generation bioethanol production: An insight to pretreatment, saccharification and fermentation processes. Renewable and Sustainable Energy Reviews. 2017;**80**:330-340

[37] Chandel AK, Chandrasekhar G, Silva MB, da Silva SS. The realm of cellulases in biorefinery development. Critical Reviews in Biotechnology 2012;**32**(3):187-202

[38] Gao Y, Xu J, Yuan Z, Zhang Y, Liang C, Liu Y. Ethanol production from high solids loading of alkali-pretreated sugarcane bagasse with an SSF process. BioResources 2014;**9**:3466-3479

[39] Öhgren K, Bura R, Lesnicki G, Saddler J, Zacchi G. A comparison between simultaneous saccharification and fermentation and separate hydrolysis and fermentation using steam-pretreated corn Stover. Process Biochemistry. 2007;**42**:834-839

[40] Bhalla A, Bansal N, Kumar S, Bischoff KM, Sani RK. Improved lignocellulose conversion to biofuels with thermophilic bacteria and thermostable enzymes. Bioresource Technology. 2013;**128**:751-759

[41] Kumagai A, Kawamura S, Lee SH, Endo T, Rodriguez JM, Mielenz JR. Simultaneous saccharification and fermentation and a consolidated bioprocessing for Hinoki cypress and Eucalyptus after fibrillation by steam and subsequent wet-disk milling. Bioresource Technology. 2014;**162**:89-95

[42] Kumar S, Dheeran P, Surendra P, Mishra IM, Adhikari D. Cooling system economy in ethanol production using thermotolerant yeast Kluyveromyces sp. IIPE453. African Journal of Microbiology Research. 2013;**1**(3):39-44

[43] Abdel-Banat BM, Hoshida H, Ano A, Nonklang S, Akada R. High-temperature fermentation: How can processes for ethanol production at high temperatures become superior to the traditional process using mesophilic yeast? Applied Microbiology and Biotechnology. 2010;**85**:861-867

[44] Radecka D, Mukherjee V, Mateo RQ, Stojiljkovic Q, Foulquie-Moreno MR, Thevelein R. Looking beyond Saccharomyces: The potential of non-conventional yeast species for desirable traits in bioethanol fermentation. FEMS Yeast Research. 2015;**15**(6):1-13

[45] Bollók M, Réczey K, Zacchi G. Simultaneous saccharification and fermentation of steam-pretreated spruce to ethanol. Applied Biochemistry and Biotechnology. 2000;**86**:69-80

[46] Koedrith PE, Dubois B, Scherens E, Jacobs C, Boonchird F. Messenguy. Identification and characterization of a thermotolerant yeast strain isolated from banana leaves. Science Asia. 2014;**34**:147-152

[47] Mejía-Barajas JA, Montoya-Pérez R, Cortés-Rojo C, Saavedra-Molina A. Levaduras Termotolerantes: Aplicaciones Industriales, Estrés Oxidativo y Respuesta Antioxidante. Information Technology. 2016;**27**(4):3-16

[48] Fonseca GG, Heinzle E, Wittmann C, Gombert AK. The yeast Kluyveromyces marxianus and its biotechnological potential. Applied Microbiology and Biotechnology. 2008;**79**(3):339-354

[49] Groeneveld P, Stouthamer AH, Westerhoff HV. Super life—How and why 'cell selection' leads to the fastest-growing eukaryote. The FEBS Journal. 2009;**276**:254-270

[50] Zhang M, Shukla P, Ayyachamy M, Permaul K, Singh S. Improved bioethanol production through simultaneous saccharification and fermentation of lignocellulosic agricultural wastes by Kluyveromyces marxianus 6556. World Journal of Microbiology and Biotechnology. 2010;**26**:1041-1046

[51] Rodrussamee N, Lertwattanasakul N, Hirata K, Suprayogi LS, Kosaka T, Yamada M. Growth and ethanol fermentation ability on hexose and pentose sugars and glucose effect under various conditions in thermotolerant yeast Kluyveromyces marxianus. Applied Microbiology and Biotechnology. 2011;**90**:1573-1586

[52] De Souza CJ, Costa DA, Rodrigues MQ, dos Santos AF, Lopes MR, Abrantes AB, dos Santos CP, Silveira WB, Passos FM, Fietto LG. The influence of presaccharification, fermentation temperature and yeast strain on ethanol production from sugarcane bagasse. Bioresource Technology 2010;**109**:63-69

[53] Jeong H, Lee DH, Kim SH, Kim HJ, Lee K, Song JY. Genome sequence of the thermotolerant yeast Kluyveromyces marxianus var. marxianus KCTC17555. Eukaryotic Cell. 2012;**11**:1584-1585

[54] Suzuki T, Hoshino T, Matsushika A. Draft genome sequence of Kluyveromyces marxianus strain DMB1, isolated from sugarcane bagasse hydrolysate. Genome Announcements. 2016;**4**(5):00923-00916

[55] Inokuma K, Ishii J, Hara KY, Mochizuki M, Hasunuma T, Kondo A. Complete genome sequence of Kluyveromyces marxianus strain NBRC1777—A nonconventional thermo-tolerant yeast. Genome Announcements. 2015;**3**(2):389-341

[56] Lertwattanasakul N, Kosaka T, Hosoyama A, Suzuki Y, Rodrussamee N, Matsutani M, Murata M, Fujimoto N, Tsuchikane K, Limtong S, Fujita N, Yamada M. Genetic basis of the highly efficient yeast Kluyveromyces marxianus: Complete genome sequence and transcriptome analyses. Biotechnology for Biofuels. 2015;**18**(8):47-61

[57] Yang C, Hu S, Zhu S, Wang D, Gao X, Hong J. Characterizing yeast promoters used in Kluyveromyces marxianus. World Journal of Microbiology and Biotechnology. 2015;**31**(10):1641-1646

[58] Zhang J, Zhang B, Wang D, Gao X. Xylitol production at high temperature by engineered Kluyveromyces marxianus. Bioresource Technology. 2014;**152**:192-201

[59] Gombert AK, Madeira JV, Cerdán ME, González-Siso MI. Kluyveromyces marxianus as a host for heterologous protein synthesis. Applied Microbiology and Biotechnology. 2016;**100**(14):6193-6208

[60] Li H, Wu M, Xu L, Hou J, Guo T, Bao X, Shen Y. Evaluation of industrial Saccharomyces cerevisiae strains as the chassis cell for second-generation bioethanol production. Microbial Biotechnology. 2015;**8**(2):266-274

[61] Signori L, Passolunghi S, Ruohonen L, Porro D, Branduardi P. Effect of oxygenation and temperature on glucose-xylose fermentation in Kluyveromyces marxianus CBS712 strain. Microbial Cell Factories. 2014;**13**:51-64

[62] Li X, Park A, Estrela R, Kim SR, Jin YS, Cate JHD. Comparison of xylose fermentation by two high-performance engineered strains of Saccharomyces cerevisiae. Biotechnology Reports. 2016;**9**:53-56

[63] Nogueira DM, Branco VCR, Moreira JRA, Pepe LMM, Goncalves FAT. Xylose fermentation by Saccharomyces cerevisiae: Challenges and prospects. International Journal of Molecular Sciences. 2016;**17**:207-225

[64] Nitiyon S, Keo-oudone C, Murata M, Lertwattanasakul N, Limtong S, Kosaka T, Yamada M. Efficient conversion of xylose to ethanol by stress-tolerant Kluyveromyces marxianus BUNL-21. Springerplus. 2016;**5**:185-197

[65] López-Alvarez A, Díaz-Pérez A, Sosa-Aguirre C, Macías-Rodríguez L, Campos-García J. Ethanol yield and volatile compound content in fermentation of agave must by Kluyveromyces marxianus UMPe-1 comparing with Saccharomyces cerevisiae baker's yeast used in tequila production. Journal of Bioscience and Bioengineering. 2012;**113**:614-618

[66] Lyubomirov SI, Yankova VP, Ivanova EP, Vengelova AK. Ehrlich pathway study in Saccharomyces and Kluyveromyces yeasts. Biotechnology & Biotechnology. 2013;**27**(5): 4103-4110

[67] Kuloyo OO, du Preez JC, García-Aparicio MP, Kilian SG, Steyn L, Görgens J. Opuntia ficus-indica cladodes as feedstock for ethanol production by Kluyveromyces marxianus and Saccharomyces cerevisiae. World Journal of Microbiology and Biotechnology 2014;**30**:3173-3183

[68] Suryawati L, Wilkins MR, Bellmer DD, Huhnke RL, Maness NO, Banat IM. Simultaneous saccharification and fermentation of Kanlow switchgrass pretreated by hydrothermolysis using Kluyveromyces marxianus IMB4. Biotechnology and Bioengineering. 2008;**101**:894-902

[69] Yan J, Wei Z, Wanga Q, He M, Li S, Irbis C. Bioethanol production from sodium hydroxide/hydrogen peroxide pretreated water hyacinth via simultaneous saccharification and fermentation with a newly isolated thermotolerant Kluyveromyces marxianus strain. Bioresource Technology. 2015;**193**:103-109

[70] Shanmugam S, Balasubramaniyan B, Jayaraman J, Ramanujam PK, Gurunathan B. Simultaneous Saccharification and fermentation of bioethanol from softwood Moringa Oleifera using thermo-tolerant yeast Kluyveromyces marxianus MTCC 1388. International Journal of ChemTech Research. 2014;**6**(12):5118-5124

[71] Cunha ACR y Inês CR. Selection of a Thermotolerant Kluyveromyces marxianus strain with potential application for cellulosic ethanol production by simultaneous Saccharification and fermentation. Applied Biochemistry and Biotechnology 2014;**172**: 1553-1564

[72] Costa DA, Souza CJA, Costa PS, Rodrigues MQRB, Santos AF, Lopes MR, Genier HLA, Silveira WB, Fietto LG. Physiological characterization of thermotolerant yeast for cellulosic ethanol production. Applied Microbiology and Biotechnology. 2014;**98**:3829-3840

[73] Yu-Sheng L, Wen-Chien L, Kow-Jen D, Yen-Han L. Ethanol production by simultaneous saccharification and fermentation in rotary drum reactor using thermotolerant Kluveromyces marxianus. Applied Energy. 2013;**105**:389-394

[74] Moreno AD, Ibarra D, Ballesteros I, González A, Ballesteros M. Comparing cell viability and ethanol fermentation of the thermotolerant yeast Kluyveromyces marxianus and Saccharomyces cerevisiae on steam-exploded biomass treated with laccase. Bioresource Technology. 2013;**135**:239-245

[75] Hyun-Woo K, Yule K, Seung-Wook K, Gi-Wook C. Cellulosic ethanol production on temperature-shift simultaneous saccharification and fermentation using the thermostable yeast Kluyveromyces marxianus CHY1612. Bioprocess and Biosystems Engineering. 2011;**35**:115-122

[76] Pessani NK, Atiyeh HK, Wilkins MR, Bellmer DD, Banat IM. Simultaneous saccharification and fermentation of Kanlow switch grass by thermotolerant Kluyveromyces

marxianus IMB3: The effect of enzyme loading, temperature and higher solid loadings. Bioresource Technology. 2011;**102**:10618-10624

[77] García-Aparicio MP, Oliva JM, Manzanares P, Ballesteros M, Ballesteros I, González A, Negro MJ. Second-generation ethanol production from steam exploded barley straw by Kluyveromyces marxianus CECT 10875. Fuel. 2011;**90**:1624-1630

[78] Faga BA, Wilkins MR, Banat IM. Ethanol production through simultaneous saccharification and fermentation of switchgrass using Saccharomyces cerevisiae D5A and thermotolerant Kluyveromyces marxianus IMB strains. Bioresource Technology. 2010;**101**:2273-2279

[79] Tomás-Pejó E, Oliva JM, González A, Ballesteros I, Ballesteros M. Bioethanol production from wheat straw by the thermotolerant yeast Kluyveromyces marxianus CECT 10875 in a simultaneous saccharification and fermentation fed-batch process. Fuel. 2009;**88**:2142-2147

[80] Kádár Z, Szengyel K. Réczey. Simultaneous saccharification and fermentation (SSF) of industrial wastes for the production of ethanol. Industrial Crops and Products. 2004;**20**:103-110

[81] Ballesteros M, Oliva JM, Negro MJ, Manzanares P, Ballesteros I. Ethanol from lignocellulosic materials by a simultaneous saccharification and fermentation process (SFS) with Kluyveromyces marxianus CECT 10875. Process Biochemistry. 2004;**39**:1843-1848

[82] Narra M, James JP, Balasubramanian V. Simultaneous saccharification and fermentation of delignified lignocellulosic biomass at high solid loadings by a newly isolated thermotolerant Kluyveromyces sp. for ethanol production. Bioresource Technology. 2015;**179**:331-338

[83] Nachaiwieng W, Lumyong S, Yoshioka K, Watanabe T, Khanongnuch C. Bioethanol production from rice husk under elevated temperature simultaneous saccharification and fermentation using Kluyveromyces marxianus CK8. Biocatalysis and Agricultural Biotechnology. 2015;**4**(4):543-549

[84] Wu WH, Hung WC, Lo KY, Chen YH, Wan HP, Cheng KC. Bioethanol production from taro waste using thermo-tolerant yeast Kluyveromyces marxianus K21. Bioresource Technology. 2016;**201**:27-32

[85] Kim S, y Dale BE. Global potential bioethanol production from wasted crops and crop residues. Biomass and Bioenergy. 2004;**26**(4):361-375

[86] Idarraga G, Ramos J, Zuniga V, Sahin T, Young R. Pulp and paper from blue Agave waste from tequila production. Journal of Agricultural and Food Chemistry. 1999;**47**:4450-4455

[87] Iñiguez CG, Bernal CJJ, Ramírez MW, Villalvazo NJ. Recycling Agave bagasse of the tequila industry. Advances in Chemical Engineering and Science. 2014;**4**:135-142

[88] Chávez-Guerrero L, Flores J, Kharissov BI. Recycling of ash from mezcal industry: A renewable source of lime. Chemosphere. 2010;**81**:633-638

[89] Mussatto SI, Dragone G, Guimarães PMR, Silva JPA, Carneiro LM, Roberto IC. Techno logical trends, global market, and challenges of bio-ethanol production. Biotechnology Advances. 2010;**28**:817-830

[90] Hernández-Salas JM, Villa-Ramírez MS, Veloz-Rendón JS, Rivera-Hernández KN, González-César RA, Plascencia-Espinosa MA. Comparative hydrolysis and fermentation of sugarcane and agave bagasse. Bioresource Technology. 2009;**100**:1238-1245

[91] Saucedo-Luna J, Castro-Montoya A, Martinez-Pacheco M, Sosa-Aguirre C, Campos-Garcia J. Efficient chemical and enzymatic saccharification of the lignocellulosic residue from Agave tequilana bagasse to produce ethanol by Pichia caribbica. Journal of Industrial Microbiology & Biotechnology. 2011;**38**:725-732

[92] Caspeta L, Caro-Bermúdez MA, Ponce-Noyola T, Martinez A. Enzymatic hydrolysis at high-solids loadings for the conversion of agave bagasse to fuel ethanol. Applied Energy. 2014;**113**:277-286

[93] Montiel C, Hernández-Meléndez O, Vivaldo-Lima E, Hernández-Luna M, Bárzana E. Enhanced bioethanol production from blue Agave bagasse in a combined extrusion-Saccharification process. Bioenergy Research. 2016;**9**(4):1005-1014

[94] Rios-González LJ, Morales-Martínez TK, Rodríguez-Flores MF, Rodríguez-De la Garza JA, Castillo-Quiroz D, Castro-Montoya AJ, Martinez A. Autohydrolysis pretreatment assessment in ethanol production from agave bagasse. Bioresource Technology. 2017;**242**: 184-190

[95] Barrera I, Amezcua-Allierib MA, Estupiñan L, Martínez T, Aburto J. Technical and economical evaluation of bioethanol production from lignocellulosic residues in Mexico: Case of sugarcane and blue agave bagasses. Chemical Engineering Research and Design. 2016;**107**:91-101

[96] Corona-González RI, Varela-Almanza KM, Arriola-Guevara E, Martínez-Gómez AJ, Pelayo-Ortiz C, Toriz G. Bagasse hydrolyzates from Agave tequilana as substrates for succinic acid production by Actinobacillus succinogenes in batch and repeated batch reactor. Bioresource Technology. 2016;**205**:15-23

Chapter 3

Renewable Energy Integration with Energy Storage Systems and Safety

Omonayo Bolufawi

Additional information is available at the end of the chapter

http://dx.doi.org/10.5772/intechopen.78351

Abstract

One of the major goals of sustainable energy systems is to provide clean, affordable, accessible energy with benign environmental impact. Development of reliable energy systems without toxic byproducts to preserve the environment while powering the future is urgently needed. This need has led to the design and implementation of power generating systems using photovoltaics (PV) and energy storage devices. Currently, there is an excess increase in the fossil fuels cost due to increase in consumption of electric grid energy and its inability to meet up with the demand. Optimizing the generation, storage and use of electric power by using renewables (PV) and storage devices will enhance efficient, effective and reliable power consumption. This chapter proposes an efficient approach for the integration of renewable energy systems (PV) and energy storage devices as well as their safety and tradeoffs in the environment.

Keywords: renewable energy, energy storage, safety, integration and grid

1. Introduction

A major goal of sustainable energy system using renewable energy is to provide clean, affordable, accessible energy with efficient energy storage with depleting the earth resources. There is a need to develop reliable energy systems that do not depend on fossil fuel to preserve the environment while powering the present and the future. This has led to the development of power generating systems utilizing renewables (photovoltaics and wind) [1]. There has being rapid increase in the power generation from PV over there the years [2]. Due to the rapid increase in PV generation, energy storage is serves as a storage medium for excess generation which can use when needed. Energy storage systems also serve as a means of increasing

the power utilization and consumption rate. Implementing battery storage is limited due to relatively high cost. In some grid connected systems, Plug in Hybrid Electric Vehicles (PHEV) is use as storage which can function as double use systems [1, 3–8]. Future smart and micro grids could benefit from the double use functionality of electric vehicles as part of the energy network to provide vehicle-to-grid services (V2G) as described in [1]. Research shows that the average car is parked 15 hours a day and can provide storage and or grid services 60% of the time [1]. The idea of integration renewable energy (photovoltaics) with energy storage devices for a double use is illustrated.

2. Systems, integration and load layout

The system block diagram shown in **Figure 1** consist of renewable energy source (RES), power and energy management system (PEMS), grid, energy storage devices (ESD), residential load. The system is design to integrate RES and ESD using the PEMS.

2.1. Renewable energy source (RES)

The renewable energy source in the design above is solar photovoltaics (PV) use for power generation. Solar cells also called PV convert sunlight directly to electricity. The power generated from solar highly depend on the amount of sun light. Maximum generation is usually achieved during peaks of day light, if sufficient enough, excess generation is stored in energy storage devices such as lithium ion batteries for later use during off peak hours usually in the early morning and late in the evening to mid-night. Considering a renewable source rated at 3 kW, roof mounted system with the area of the solar cells approximately 20 m^2 with an efficiency of 15%. The power from the PV system is determined using a linear model based on the irradiance level. The equation representing the simplified model is given in [4]:

$$PV_{output(t)} = GHI(t) \times S \times PV_{\eta} \tag{1}$$

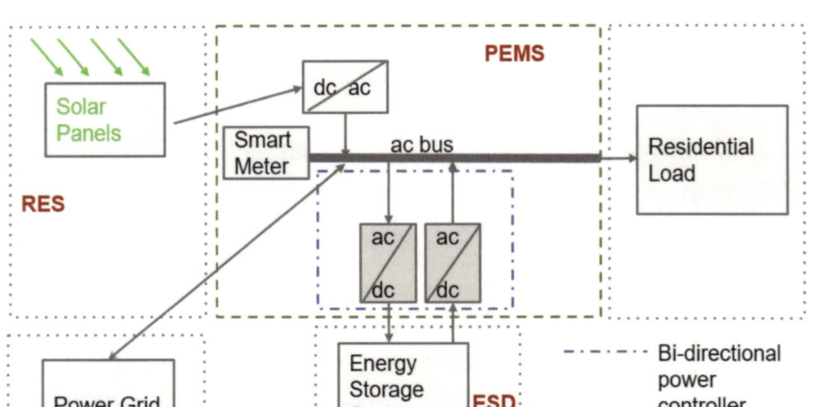

Figure 1. Detailed diagram of the integration module.

where GHI(t) is the global horizontal irradiation in W/m^2, S is the total area for the PV modules in m^2 and PVη is the efficiency of the PV modules.

The PV generator is connected to the system via a DC to AC inverter with maximum power point tracking and constant efficiency. The change in efficiency of the inverter depending on the input and required output are not considered. The generation of electricity for PV is also temperature sensitive and that is also not considered in this project. This however does not significantly hamper the system and this simplification has been applied successfully in previous works including [4, 9].

The PV system was implemented and simulated using the System Advisor Model (SAM) developed and distributed by the National Renewable Energy Lab (NREL) [9]. A detailed residential PV model was developed. The PV generator size was chosen to cover our peak load requirement of 2.5 kW while being reasonably priced and having a footprint that can easily fit on the rooftop of an average-sized house. The SAM simulation, using the GHI data for the south side of Tallahassee, provided the expected power production from the PV generator for the period of a year. This is shown in **Figure 2** along with the expected AC production.

2.2. The power and energy management system (PEMS)

The PEMS in the system design serve as control for energy flow and conversion from RES, grid and energy storage. It contains the power electronics required to interface the power systems and the load. The PV generator is connected to the AC bus via an inverter and the energy storage device is connected the AC bus via a bi-directional controller as shown if **Figure 1**. Net

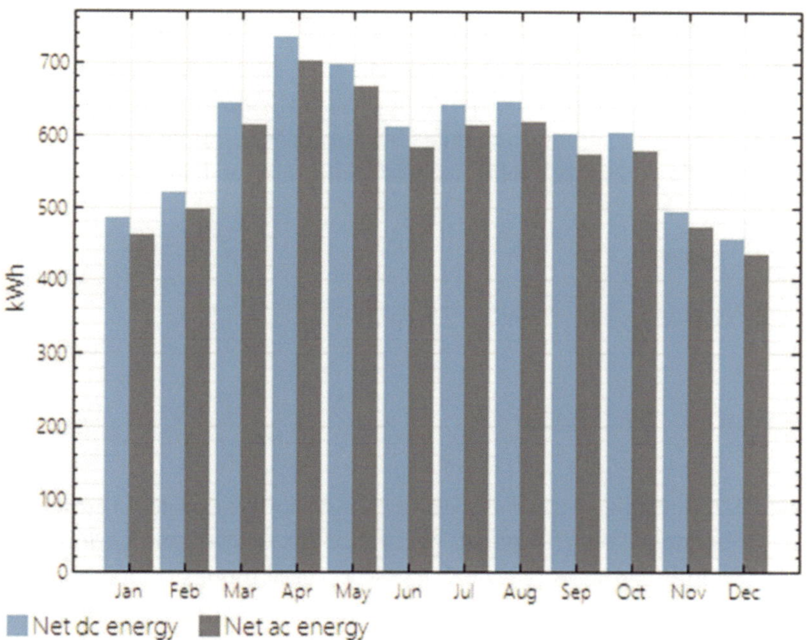

Figure 2. PV production over the period of 1 year based on SAM simulation [9].

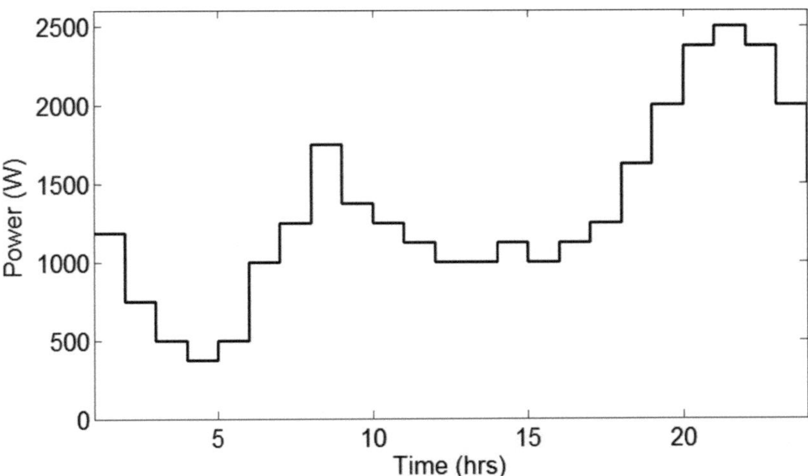

Figure 3. Residential load for system analysis with hourly resolution.

metering is in effect to account for excess flow of power to the grid and the meter is smart to monitor the power usage of different components. Two DC-DC boost converters and a single inverter is use in the power and energy management system. This is due to the differences in the output of the PV and the storage device voltage. This PV array converter also function as maximum power point tracking (MPPT). The power flow in the circuit is been monitored and controlled by the operation mode control logic which is embedded in the PEMS.

2.3. Energy storage devices (ESD)

Energy storage devices plays a vital role in the transition to clean, efficient and reliable power source for energy sustainability. In this chapter, lithium ion battery (LIB) is used as the ESD. LIBs are highly in demand for portable electrical/electronic devices and commercial application. It is currently gaining traction as backup power source for residential. This follows the lunch of Tesla power wall which consist of lithium ion batteries.

2.4. System load

A typical residential house load is illustrated in **Figure 3**.

3. RES integration with ESD

The integration of renewable energy source and energy storage devices is growing immensely to reduce overdependence on grid power generation. In this section, various mode of integration principles and operation will be discussed. These modes of operations account for different conditions that can affects the RES, ESD and Grid during a 24 hours' period which observes the morning, afternoon, evening and night times.

3.1. Operation mode of grid-supported PV

This mode of operation topology is showed in **Figure 4**. The figure describes the cases when the PV is generating but not sufficient to power the residential load. The grid is used to supplement the required power for the residential load. The ESD used due to cases of low state of charge or off-peak grid price. During this period of off peak grid pricing, the ESD also can be charge and not discharge because of the losses due to the round-trip efficiency from the AC bus to the battery and back to the AC bus in **Figure 1**.

3.2. Operation mode of grid and energy storage system-supported PV

The mode of operation describes cases when the PV generation is below the residential load requirement and need to supplement with both power from the energy storage and the grid before meeting the load required. **Figure 5** illustrates this mode of operation.

3.3. Operation mode of energy storage system-supported PV

This mode describes when the PV power is not sufficient to power the load and the ESD is available. In this scenario, the load is power by the PV and the ESD. This mainly happens during on-peak grid pricing times when the ESD is charged during off-peak grid pricing times (**Figure 6**).

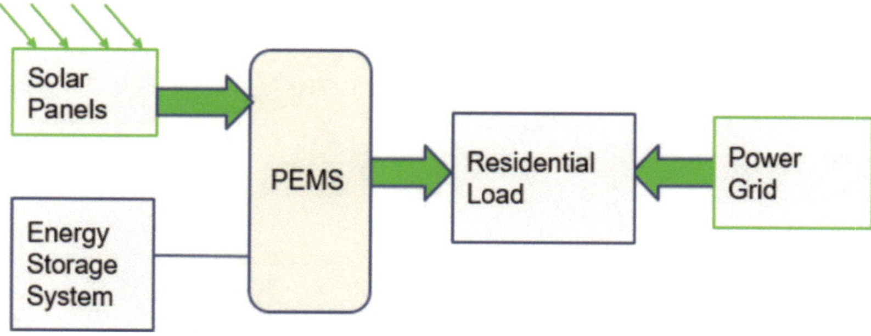

Figure 4. Grid and PV supplying the residential load.

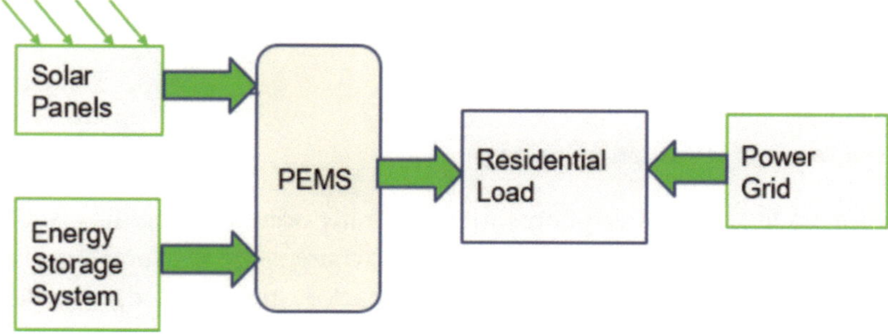

Figure 5. The grid, ESD and PV supplying the residential load.

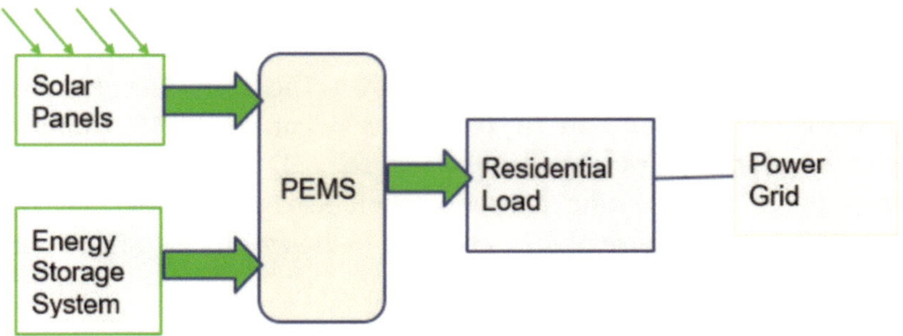

Figure 6. The PV and ESD supplying the residential load.

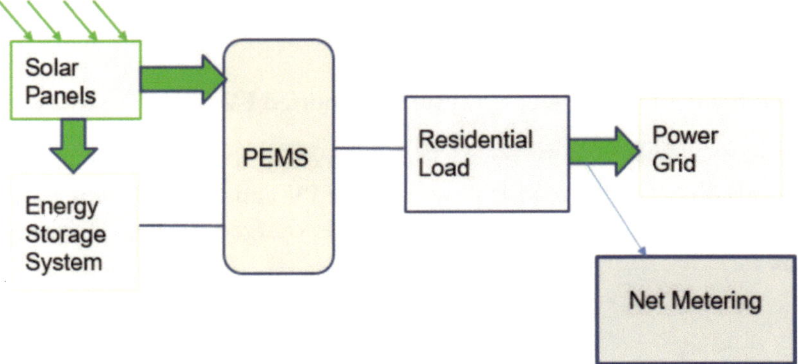

Figure 7. The PV supplying the residential load, grid and charging the ESD.

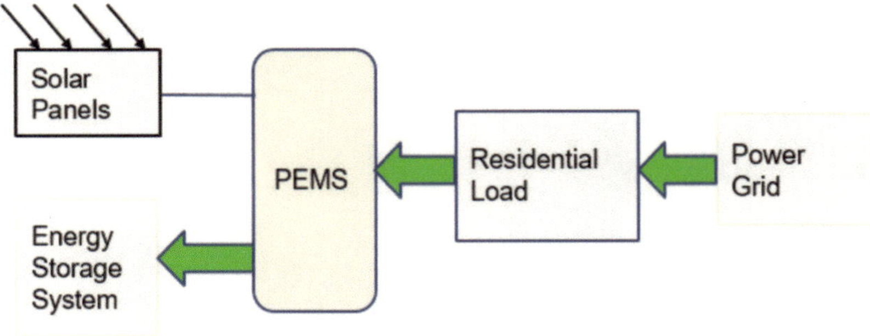

Figure 8. The grid supplying the residential and charging the ESD.

3.4. Operation mode of PV in peak hours

Figure 7 shows mode 4, this mode of operation generally occurs during the peak hours of PV generation. In this scenario, the PV powers the load, charges the ESD and supplies power to the grid for credit. The process of supplying power back to grid is called net metering.

3.5. Operation mode of grid

This mode of operation describes cases when the PV is not generating any power. The power needed by the load is supplied from the grid. The grid also charges the ESD when the state of charge is less than the thresh hold (maximum charge). Charging is usually done on off-peak grid hours to save cost (**Figure 8**).

4. Safety issues relating with the integration of RES and ESD

To effectively have a safe and reliable sustainable energy systems. Safety is highly imperative in the integration of the RES and ESD. Energy storage devices (ESD) such as lithium ion battery a high-performance storage device is used but has a drawback in its safety based on their material and chemical composition. Lithium ion batteries are the enabling technology for storage solutions in many applications. A typical Li-ion cell consists; positive electrode, negative electrode, electrolyte, and separator. Like all electrochemical batteries, the chemical energy is converted into electrical energy. According to the second law of thermodynamics, any conversion between two forms of energy occurs with an energy loss. This energy loss increases the temperature of the cell, which, negatively, affects the battery life, and possibly exceeds safety limits. The safety issues related to Li-ion batteries are caused by abuse conditions which are basically divide into electrical, mechanical and environmental, can result in abrupt behavior of the batteries. The electrical abuse conditions include, short circuiting and overcharge with an over-charging current rate and charging time. The mechanical abuse conditions include crush test, nail penetration test and external heating. These off-nominal operating conditions can lead to critical failure of lithium ion batteries. Safety of lithium ion battery has been the technical obstacle for high power demand applications in RES and ESD integration, hybrid electric vehicles (HEV) and electric vehicles (EV). These abuse conditions can initiate thermal runaway in LIB wherein chain exothermic reaction can cause the battery to attain temperature of over >500°C. The major response of the cells to the abuse conditions is usually increased in temperature due to decomposition of the electrolyte, melting of separator which in turn leads to exothermic reaction. Most of these responses eventually leads to thermal runaway before resulting into explosion or fire. The safety of LIB directly related to the type of material chemistry and their thermal stability.

The unforeseen battery failure potential has created public awareness for battery safety, particularly because of very large product recalls involving battery failures. Typical battery failure response can be energetic and non-energetic. Both energetic and non-energetic failures of Li-ion batteries often occur for due to poor cell design (electrochemical or mechanical), cell manufacturing flaws, external abuse of cells (thermal, mechanical, or electrical), poor battery pack design or manufacture, poor protection electronics design or manufacture [10]. Thus, Li-ion battery reliability and safety are generally considered a function of the design and manufacturing process. Standard performance regulation has been designed to test cell and battery pack designs to pass compliance with Underwriter Laboratories (UL), United Nations (UN) organizations standards. Failures that occur in the field are seldom related to cell design; rather, they are predominantly the result of manufacturing defects or subtle abuse scenarios that result in the development of latent cell internal faults. The failure modes can be classified into Energetic and Non-energetic failures.

- Energetic failure: This type of failures often leads to thermal runaway. Thermal runaway refers to excessive heating of a cell due to exothermic chemical reaction, this is a major occurrence with batteries when abused [10]. During this failure, the energy stored in the cell is rapidly removed. The rate of thermal runaway is proportional to the amount of energy stored. The thermal runaway reactions level is also related to the cell content such as electrolyte and electrode material chemistry [11].

- Non-Energetic failure: This type of failure mode which is usually considered benign result to loss of capacity, internal impedance increase. The ideal lithium ion battery failure mode is slow capacity fading and internal impedance increase caused by normal aging of the cells within the battery [11].

4.1. Temperature hazard of a lithium ion battery

Temperature is a major hazard response of lithium ion battery when subject to an abuse condition. Most lithium ion batteries should operate in temperature range of −10 to 50°C. During low temperature, reactions rate is slower according to Arrhenius law, which also reduce the transfer rate of ions and electron and causes reduction in the capacity. Lithium ion batteries at high temperature, are more vulnerable to high risk of failure than at lower temperature. Major failure response resulting in destruction of batteries are due to relatively high temperature. This could be because of electrolyte decomposition, melting of separator which in turns leads to thermal runaway after exothermal reactions before eventually resulting into explosion and/or fire. Thermal runaway is an adverse condition that is caused by a battery charging or other process that produces more internal heat than it can dissipate which refers to a situation where an increase in temperature changes the condition of the battery that further increases the temperature, often lead to explosion and/or fire. The type of abuse condition and the cell chemistry as well as the design affect the cell reaction. The onset temperature of the thermal runaway depends on the chemistry of the battery. This reaction is sustained by battery's oxygen content which varies by different cathode materials. The occurrence of a cell thermal runaway event depend on factors including; the state of charge of a cell (volume of electrical energy stored in the form of chemical potential energy), the ambient temperature, the cell electrochemical design (cell chemistry), and the mechanical design of the cell (cell size, electrolyte volume, etc.) [12]. The most severe thermal runaway reaction will be achieved when the cell is at 100% SOC, or subject to an abuse conditions such as overcharge, short circuiting, crushing etc. The following occurs when a cell undergoes thermal runaway;

- Cell internal temperature increases.
- Cell internal pressure increases
- Cell undergoes venting
- Cell content may be ejected.

The general root cause of energetic battery failure is; electrical abuse, mechanical abuse, poor electrochemical design and thermal abuse. Each of these failure modes have an impact on the environment.

4.2. Thermal characterization of Li-ion battery

The lithium ion battery relatively has large volume with a small surface area which makes the extraction of heat from the battery very important. There is an increase in the battery temperature if the heat generated is not removed which can lead to thermal runaway. Abuse by overcharge adds energy to the battery due to the input of electric power while in short circuit test no energy is added to the cell [13]. This could also affect the amount of energy delivered by the battery. To get optimum performance and effectively maximize the battery, it is imperative to operate the battery within the specify temperature range by the battery's manufacturer [15]. For instance, the preferred operating temperature for most lithium ion battery is −20 to 50°C. The second law of thermodynamics limits the rate of energy conversion during charging and discharging, leading to a non-ideal process with energy loss in the form of heat [14]. Bernadi et al. [14] classic work, estimates the heat generated from batteries using a mathematical model. According to their study, the heat is generated due to four main reasons; the irreversible resistive heating, the reversible entropic heat, the heat change of chemical side reactions, and heat of mixing due to the generation and relaxation of concentration gradients. The irreversible resistive (ohmic loss) occurs during both charge and discharge when the battery current flows through an internal resistance. The irreversible resistive heat occurs following the deviation of the battery potential from its equilibrium potential due to internal resistances. Therefore, the differences between the terminal voltage and the open circuit voltage is converted into heat [15].

The reversible entropic heat is the heat absorbed by the battery itself through a temperature change. Heat generated from the battery can also be determined experimentally by methods such as temperature measurements, thermal imaging, and calorimeter. A Calorimeter is a device that measures the amount of heat released and/or absorbed during a process. Differential Scanning Calorimetry (DSC) provides method of determining thermal stability by an induced heat and the subsequent heat generated by different materials is measured.

5. Conclusion

This chapter discussed safe integration of renewable energy with energy storage devices which is needed to have a reliable and efficient sustainable energy systems. Proper implementation of the different modes of operation which considers the working state of RES, ESD and grid will immensely reduce the over dependence on grid especially during on peak grid pricing. The ever increasing environmental problem will reduce drastically when renewable source of power is used with adequate storage capacity for energy sustainability. Improving the energy storage device capacity is necessary but also poses safety risk as well because the higher the capacity of energy storage device, the higher the safety risk associated with it. Therefore, there is need to effectively balance these tradeoffs in order to have both safe and high performing systems.

Author details

Omonayo Bolufawi

Address all correspondence to: omonayo1.bolufawi@famu.edu

Florida A&M University, Tallahassee, United States

References

[1] Cvetkovic I, Thacker T, Dong D, Francis G, Podosinov V, Boroyevich D, Wang F, Burgos R, Skutt G, Lesko J. Future home uninterruptible renewable energy system with vehicle-to-grid technology. In: Energy Conversion Congress and Exposition, 2009. ECCE 2009. IEEE; 20-24 Sep 2009. pp. 2675-2681. DOI: 10.1109/ECCE.2009.5316064

[2] U.S. Energy Information Administration (EIA). Available from: www.eia.gov

[3] Jayasekara N, Wolfs P. A hybrid approach based on GA and direct search for periodic optimization of finely distributed storage. In: Innovative Smart Grid Technologies Asia (ISGT), 2011 IEEE PES. 13-16 Nov 2011. pp. 1-8. DOI: 10.1109/ISGT-Asia.2011.6167137

[4] Yu R, Kleissl J, Martinez S. Storage size determination for grid-connected photovoltaic systems. IEEE Transactions on Sustainable Energy. Jan 2013;**4**(1):68-81. DOI: 10.1109/TSTE.2012.2199339

[5] Goli P, Shireen W. PV integrated smart charging of PHEVs based on DC link voltage sensing. IEEE Transactions on Smart Grid. May 2014;**5**(3):1421-1428. DOI: 10.1109/TSG.2013.2286745

[6] Nagarajan A, Shireen W. Grid connected residential photovoltaic energy systems with plug-in hybrid electric vehicles (PHEV) as energy storage. In: 2010 IEEE Power and Energy Society General Meeting. 25-29 July 2010. pp. 1-5. DOI: 10.1109/PES.2010.5589387

[7] Chen C, Duan S. Optimal integration of plug-in hybrid electric vehicles in microgrids. IEEE Transactions on Industrial Informatics. Aug 2014;**10**(3):1917-1926. DOI: 10.1109/TII.2014.2322822

[8] Gitizadeh M, Fakharzadegan H. Effects of electricity tariffs on optimal battery energy storage sizing in residential PV/storage systems. In: 2013 International Conference on Energy Efficient Technologies for Sustainability (ICEETS). 10-12 Apr 2013. pp. 1072-1077. DOI: 10.1109/ICEETS.2013.6533536

[9] National Renewable Energy Lab. Available from: www.nrel.gov

[10] http://www.sfpe.org/page/2012_Q4_2/Lithium-Ion-Battery-Hazards.htm

[11] http://www.baj.or.jp/e/knowledge/structure.html

[12] Mikolajczak C, Kahn M, White K, Long RT. Lithium Ion Batteries Hazard and Use Assessment. Exponent Failure Analysis Associates, Inc. July 2011 (Fire protection research foundation)

[13] Larsson F, Mellander B-E. Abuse by external heating, overcharge and short circuiting of commercial Lithium-ion battery cells. Journal of the Electrochemical Society. 2014;**161**(10):A1611-A1617. DOI: 10.1149/2.0311410jes

[14] Bernardi D, Pawlikowski E, Newman J. A general energy balance for battery systems. Journal of the Electrochemical Society. 1985;**32**(1):5-12. DOI: 10.1149/1.2113792

[15] Jaiswal A. Lithium-ion battery based renewable energy solution for off-grid electricity: A techno-economic analysis. Renewable and Sustainable Energy Reviews. 2017;**72**:922-934. DOI: 10.1016/j.rser.2017.01.049

Chapter 4

Microgrid Integration

Mulualem T. Yeshalem and Baseem Khan

Additional information is available at the end of the chapter

http://dx.doi.org/10.5772/intechopen.78634

Abstract

Hybrid energy systems are becoming attractive to supply electricity to rural areas in all aspects like reliability, sustainability, and environmental concerns, and advances in renewable energy technology; especially for communities living far in areas where grid extension is difficult so generation of renewable energy resources like solar and wind energy to provide reliable power supply with improved system efficiency and significant cost reduction is best way. Besides this, the demand for renewable energy source in large urban cities is increasing, and their integration to the existing conventional grid has become more fascinating challenges. So the future requires stable and reliable integration of renewable distributed generators to the grid, and the local loads are close to distributed generators. Most existing power plants have centralized control system and remote power generation site while most renewable power generations are distributed and connected to lower or medium voltage networks near the customer. When the power demand increases, power failure and energy shortage also increase so the renewable energy can be used to provide constant and sustainable power. The chapter will provide a complete overview of microgrid system with its complete operation and control.

Keywords: distributed generation (DG), microgrid, grid integration and control, renewable energy

1. Introduction

The conventional power network comprises large generating stations with extra high voltage links, which connect transmission substations with distribution system for delivering power to end users. Therefore, the basic concept in traditional power system is the central controlling with unidirectional energy flow for transmitting power to load centers.

Renewable sources of energy are becoming the most important sources for supplying electrical energy straight to the customer without traditional distribution system, especially for communities living far in areas where grid extension is difficult so generation of green sources such as photovoltaic (PV) and wind power for providing reliable energy with improved system efficiency and significant cost reduction is best way. Therefore, renewable energy sources are the most sustainable remedies for producing energy and heat. The main advantages of renewable energy sources are the instant availability, less dependence on fossils fuels, low cost variation, and no transportation cost with higher economic efficiency [1].

Microgrid is among the new technology that has attracted a great attention; recently, the dependency on the centralized power system is changing; and replaced by smaller and more distributed generation (DG) located closer to load to meet their requirements effectively and heat. That is, everyone is a producer and consumer of energy at the same time; by doing so, they became energy independent from the overcharging utility companies. Microgrids are not a replacement for traditional utility infrastructure. From the utility viewpoint, the transmission and distribution cost is lowered; reduction in line losses, network congestion, and load shedding; improvement in power quality and reliability; and reduction in infrastructure investment needs.

Microgrid consists of group of multiple distributed sources with interconnected demands. It is operated either in stand-alone mode or grid connected mode [2, 3]. Microgrid can be defined as a system or a subsystem, which incorporates single, or multiple sources, controlled demands, energy storage systems, security and supervision system. These elements and subsystems make microgrid operational in utility integrated or isolated mode. Here, the main function of the utility grid is to maintain system frequency and bus voltage by supplying deficient power instantly [4]. Microgrid consists of bidirectional connections that means it can transmit and receive power from utility grid. Wherever any fault occurred on utility grid, microgrid switched to stand-alone mode [5]. Even though, emerging power electronic (PE) technologies and digital control systems make possible to build advanced microgrids capable to operate independently from the grid and integrating multiple distributed energy resources. There are a lot of challenges in integration, control, and operation of microgrid to whole distribution system. Microgrid is not designed to handle the large power being fed by the utility distribution feeders. Further, the characteristic of large microgrid components possesses big challenges. The issues related to the integration of microgrid raise the challenges to operation and control of main utility grid. Therefore, this chapter deals with the various microgrid integration issues faced by the main utility in the practical power system.

2. Microgrid power system

Microgrid system is a configuration of single or multiple renewable energy sources with even nonconventional sources as main energy generation source, so that the capacity shortage of power from one source will substitute by other available sources to provide sustainable power. Additionally, it incorporates energy storage and power electronics circuitry [6]. Some of the components produce direct current (DC) power and other alternating current (AC) power directly with no use of converter.

http://dx.doi.org/10.5772/intechopen.78634

2.1. Microgrid power systems configurations

Microgrid is configured based on the following technical topologies to couple the available renewable sources and to meet the required load. Here, voltage and the load demand are the determinant factors. According to [6, 7], any power system configurations are grouped in the following forms.

2.1.1. AC/DC-coupled microgrid systems

Different configurations are described in [8, 9] for the microgrid, which contains wind turbine, PV system, a diesel generator, and a battery storage system. Generally, for microgrid technological configurations, three established classes are there and are discussed below.

2.1.1.1. Microgrid systems: AC coupled

In this configuration, various renewable sources and the energy storage system are linked at the AC bus with the demands. For this type of configuration, two subcategories are available.

2.1.1.1.1. Centralized AC-coupled microgrid

All the elements are linked to the AC bus. AC power producing elements are connected to AC line in direct manner or with the help of AC/AC converter, for getting even component coupling topology. For controlling the energy flow to the battery and from the battery to the load, the master inverter required. Furthermore, DC electricity can be provided from battery if needed. **Figure 1** depicts centralized AC-coupled hybrid system configuration.

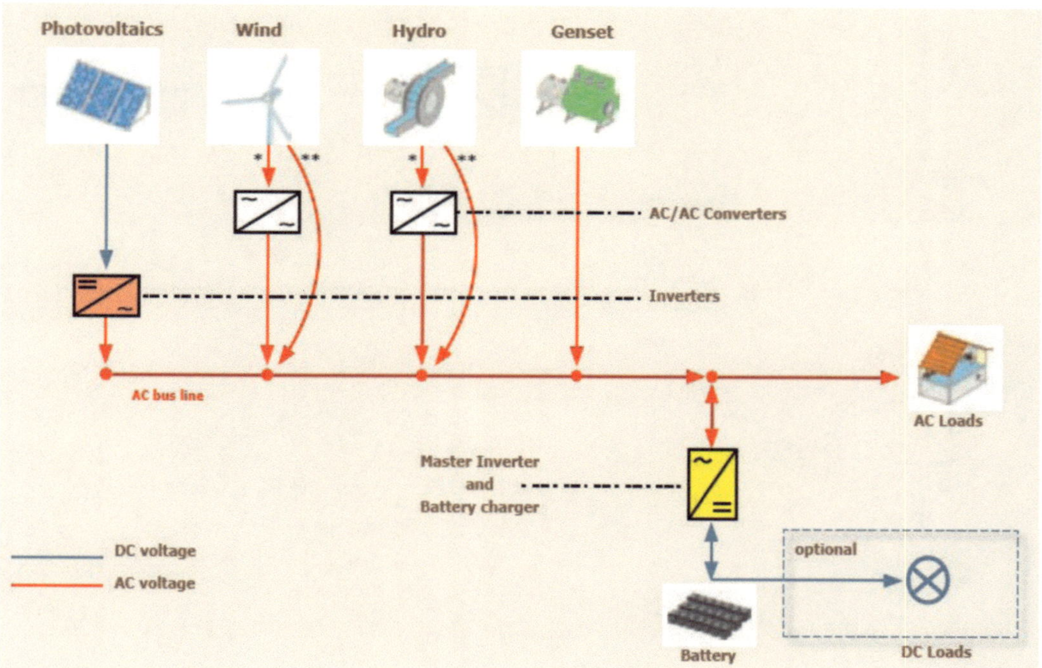

Figure 1. AC-coupled centralized microgrids [10].

2.1.1.1.2. Decentralized AC-coupled microgrid

In this type of architecture, all the technologies are not connected to any of the bus, rather they individually connect to the load directly as shown in **Figure 2**. The energy sources may not be situated in one location or close to one another and they can connect to the load from anywhere the renewable resources are available. The merit of such configuration is that the power-generating components can install from the location where renewable resource is available. But it has a disadvantage due to the difficulty of power control of the system. Thus, comparing the two configurations, the centralized system is better due to its controllability than the distributed system [6].

2.1.1.2. Microgrid system: DC coupled

In the direct current (DC) combination, all the energy sources are linked to the DC bus prior to connect the AC bus as illustrated in **Figure 3**. All AC power sources are converted to DC and then linked to the AC demand by using converter. The merit of DC-coupled topology is that the demand is met with no cutoffs. Despite the advantage of this, it has disadvantages of low conversion efficiency and no power control of diesel generator. Wind turbine and diesel generator produce AC voltage and need AC/DC converter to supply appropriate load to the

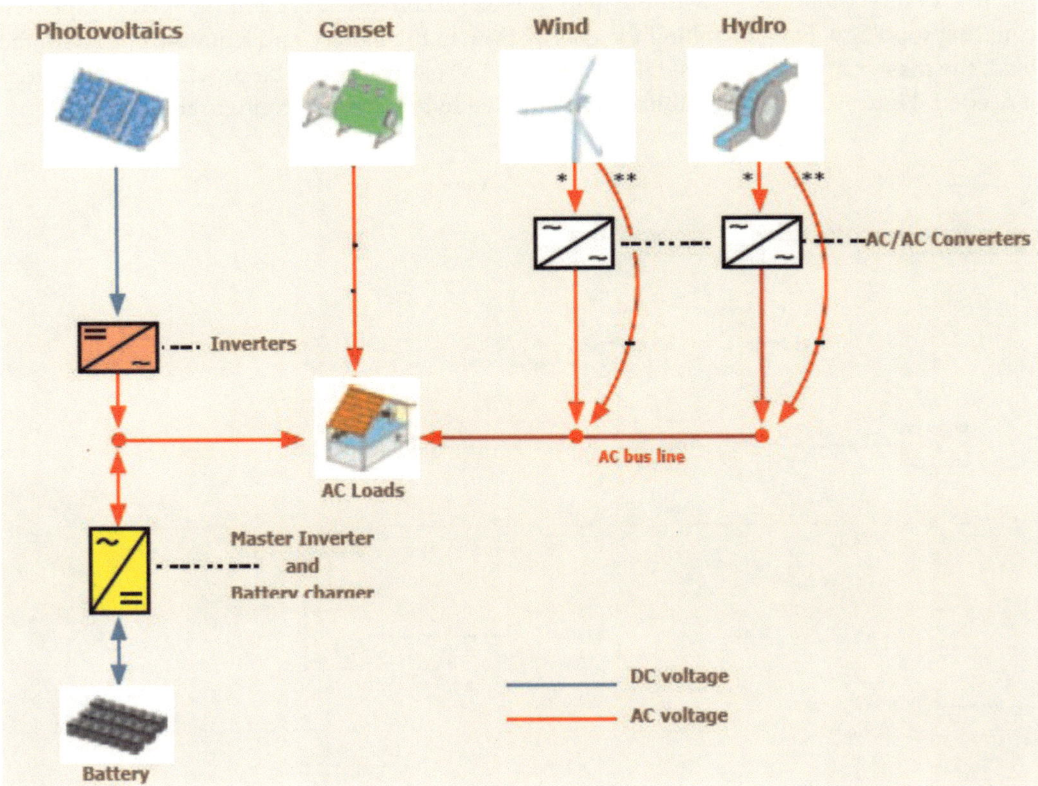

Figure 2. Decentralized AC-coupled microgrids [10].

http://dx.doi.org/10.5772/intechopen.78634

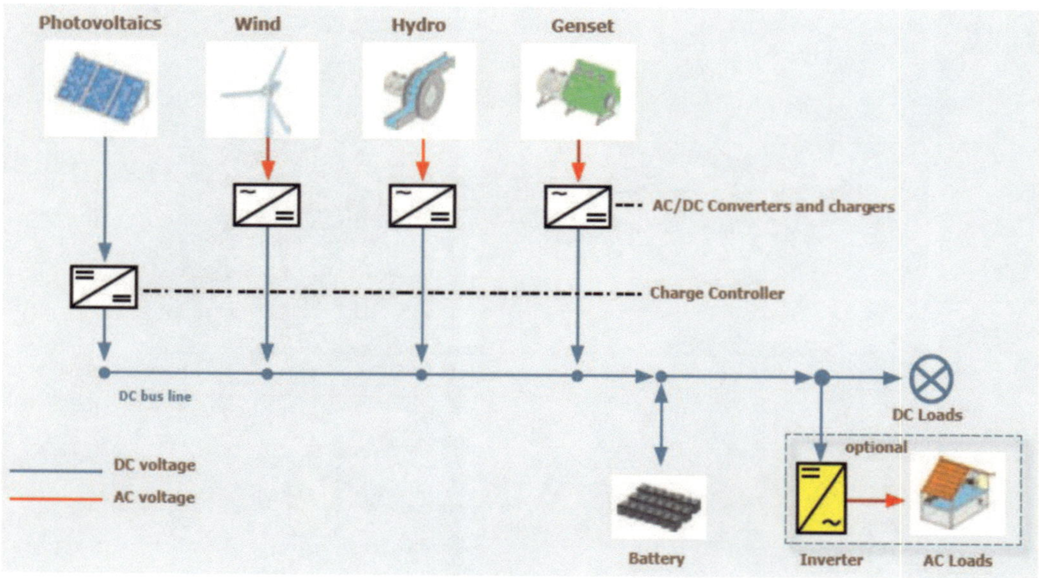

Figure 3. DC-coupled microgrids [10].

DC bus. Charge controller is also employed to protect the deep discharge and over charge of the battery. If required, AC load can be supplied using inverter.

2.1.1.3. Microgrid system: mixed coupled

There is a possibility to join AC- and DC-coupled microgrid systems. This type is called mixed-coupled microgrid system [8, 9]. In this kind of topology, some renewable are linked with battery storage at DC bus, while others are linked with DC at AC bus. **Figure 4** presents such configuration.

2.1.2. Series/parallel microgrid power system

Microgrid systems are also categorized on the basis of type of supply provided to the demands from renewables and diesel generators [6, 11]. Series and parallel hybrid microgrid are the two configurations and their detail discussions are given as follows.

2.1.2.1. Series microgrid power system

In this configuration, all the generated DC power supplied to the battery. Therefore, the energy produced by the PV, wind, and diesel generator is utilized for charging battery storage. Hence, charge controller is equipped with each component, other than diesel generator. Diesel generator is equipped with a rectifier. Afterward, inverter converts the DC power into standard AC power and feed to the AC demands. Overcharging of the battery storage by PV/wind is prevented by charge controllers. Similarly, deep discharging of the battery bank is also prevented by the charge controller. This topology is also called centralized DC bus

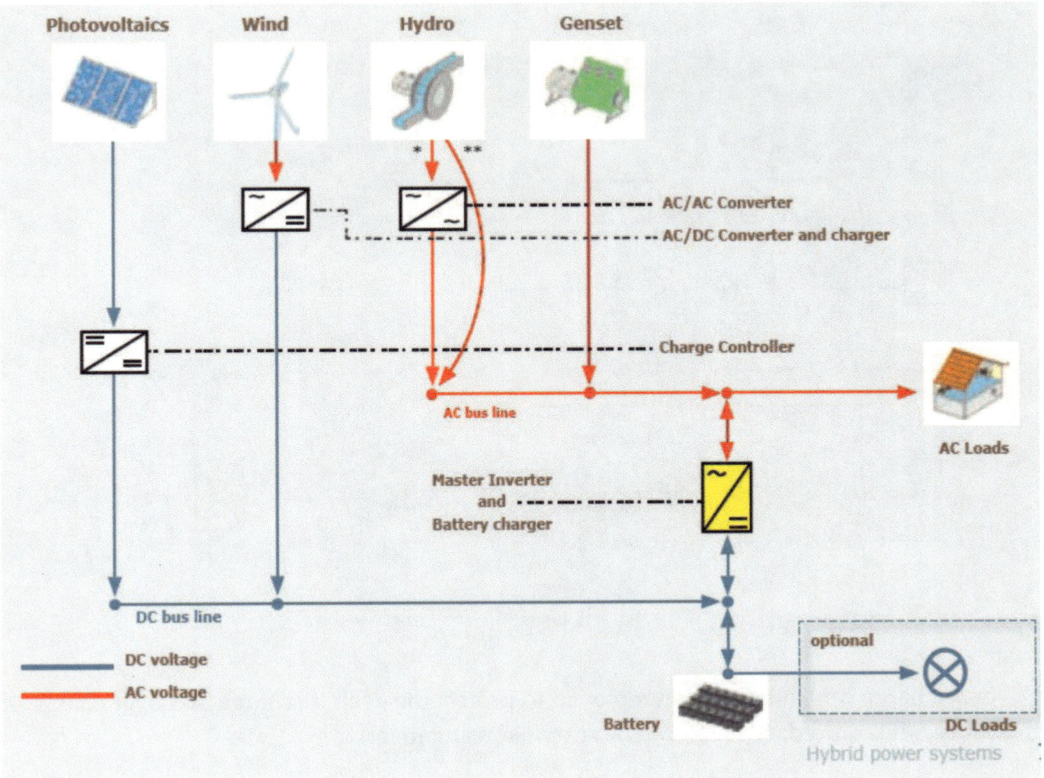

Figure 4. Mixed microgrids [10].

configuration, because all the sources are linked to DC bus and load is fed through a single point. **Figure 3** presents series microgrid power system.

2.1.2.2. Parallel microgrid power system

In this type of configuration, a part of supply demand is directly fed by the renewable sources and diesel generator directly. This configuration further classified into two subconfigurations: DC coupled and AC coupled, which are already discussed in this chapter in previous section.

3. Microgrid structure

It is a distribution network which is supplied through low and medium voltage distribution lines. Various self-sufficient and independent distributed energy sources, i.e., PV, wind, fuel cell, microhydro, etc., and storage devices such as battery storage, flywheel storage, etc., along with demands, are incorporated and grouped inside microgrid structure. **Figure 5** presents a typical overview of microgrid structure. Different distributed energy sources are integrated in microgrids by its corresponding bus bars equipped with power electronics converter. Point of common coupling (PCC) is the point where microgrid is connected to the upstream network.

There are two modes in which microgrid operates. The first one is the grid connected mode and another one is the stand-alone mode or islanded mode. In grid interfaced mode of operation, PCC is closed and microgrid is linked with utility grid. Whenever there is any disturbance in utility grid or microgrid, PCC is opened and a microgrid is disconnected to the main grid, then the microgrid is operated in stand-alone mode [11].

There are two types of microgrids available. They are AC microgrid and DC microgrid, which are depending on distributed sources and demands connected. DC grid has the advantage of easier control. Further, it does not require DC-AC or AC-DC converters; therefore, it provides lower cost and better efficiency. On the other hand, AC grid has the advantage of full utilization of available AC grid technologies but it requires synchronization and stability from the reactive power point of view [12]. **Figure 6** presents the massive DGs in the power system.

DGs provide sufficient generation to supply all, or mostly loads connected, which is linked to the microgrid. While many renewable energy systems (RES) are large scale and are connected directly to the transmission system, there are small-scale and interconnected distributed energy resources located near consumption points within low-voltage electric distribution to achieve efficient and economical requisites. DG should be provided at strategic points in the microgrid system. These strategic points may be load centers. So that, these sources provide voltage and capacity support, reduce line losses, and improve stability [14].

The renewable energy production is further classified into dispatchable and nondispatchable production. Dispatchable production is able to change their power production upon demand and by the request of grid operators. They are microhydro and megahydropower, ocean/marine current power and wave power, geothermal and ocean thermal energy conversion, biofuel biomass, etc. [15]. Nondispatchable renewable energy-based generators are wind

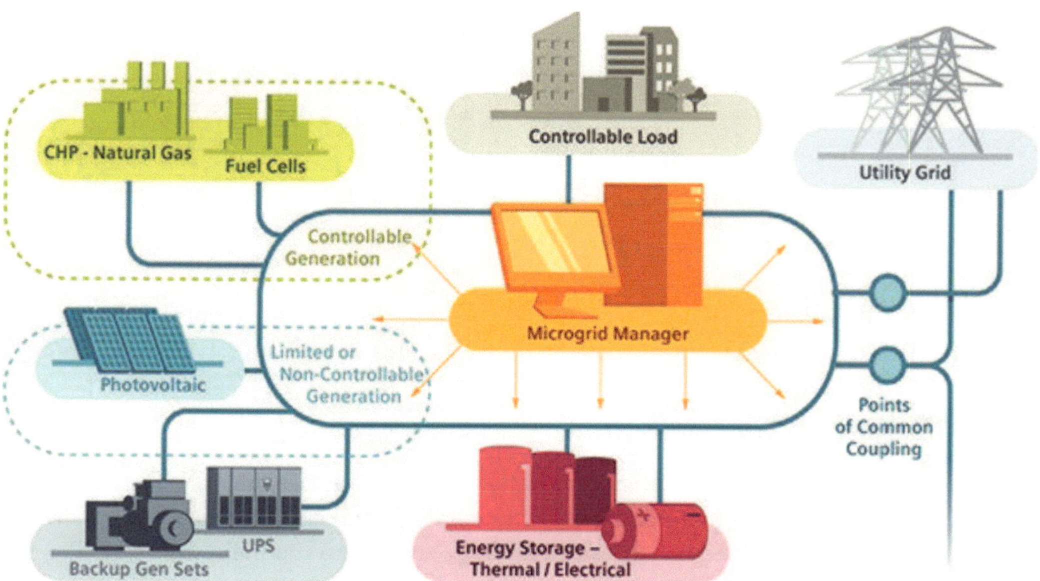

Figure 5. Microgrid power system [10].

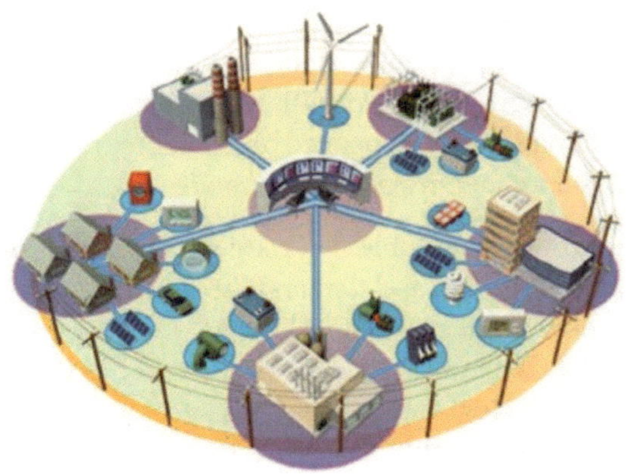

Figure 6. Massive DGs in the electrical network [13].

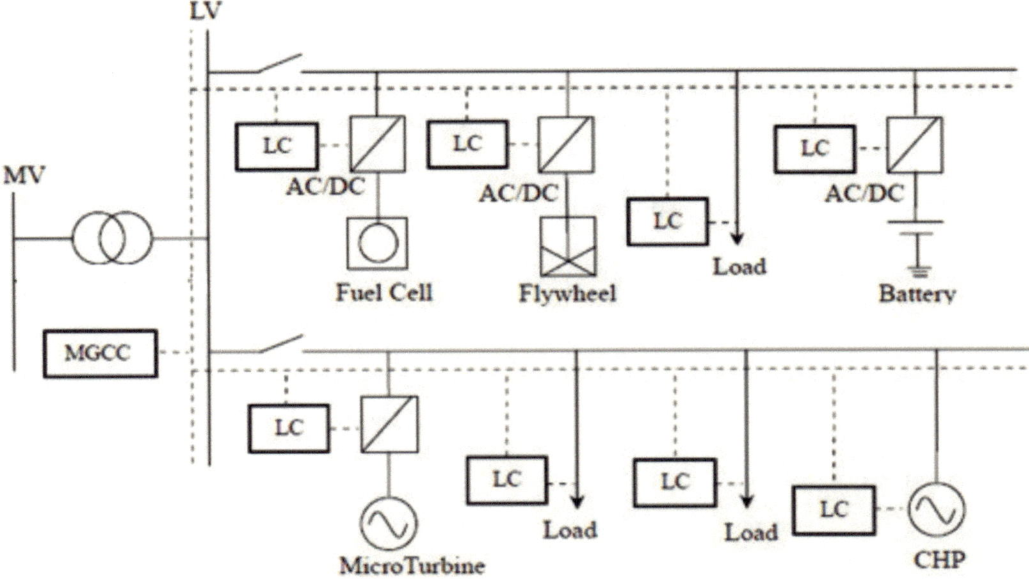

Figure 7. Basic microgrid architecture with an MGCC [17].

energy and photo voltaic, because wind turbine output depends on the wind speed and solar power available by the radiant light and heat of the sun [16].

3.1. Basic microgrid architectures

The assumed simple structure of microgrid network will be radial with several distribution feeders from different substations and a collection of loads and energy sources as illustrated in **Figure 7**. The radial power line arrangement is connected to the main utility grid at the point of common coupling (PCC) through a separation device, usually a static switch; having circuit breaker and a power flow controller for each feeder line [17]. The limited capacity of distribution generators has resulted in the development of several microgrids, which are interconnected to each other and operate with or without the main grid.

The overall architecture of a microgrid consists of an LV network on the consumer load side (both critical and noncritical loads), both noncontrollable and controllable power generators, energy storage units, and a hierarchical energy management. Controlling and monitoring each DG and loads and also managing energy system require communication infrastructure to support the control scheme so that the microgrid central control (MGCC), the center for the hierarchical control system, followed sequential low control level, like local controllers (LCs) of loads and DGs which exchange information with the MGCC for managing the whole MG operation by providing set-points to LCs. The common relevant data in exchange include mainly information about MG switch orders that are sent by the MGCC to LC and the sensed voltage/current information to MGCC from each local capture; power and frequency reference setting for each source and the state charge and discharge of the energy storage, and the protection device conditions the system in the case of fault happening to isolate the abnormal zone of the system.

3.2. Integration of microgrid to the main grid

Most of the small-scale DG sources in the load side are integrated at medium or low voltage network as low penetration fashion where they are connected as passive systems and they are not involving grid voltage controlling, frequency controlling, and stability activities. Still in the case of high penetration, the interfaces can be modified to work as active generators so that DER can participate in the frequency, voltage, and system stability control activities of the grid. Power electronic is used to interfaces between the grid and the renewable power source of microgrid so that there are not any negative influences in reliability, stability, and power quality of the supply after the interconnection DERs to the grid. Numerous components and constraints are involved in the integration of DER to the utility grid [18]. The integration of varying intermittent renewable sources like solar and wind energy conversion systems to the grid can provide a technical relief in the form of reduced losses, reduced network flows, and voltage drops. However, there are also several undesirable impacts due to high penetration of these variable DERs which include voltage swell, voltage fluctuations, reverse power flow, changes in power factor, injection of unwanted harmonics, frequency regulation issues, fault currents, and grounding issues and unintentional islanding [18]. Advanced protection system should be included in the DG units to disconnect the units in case of fault or unfavorable grid conditions.

Grid integration of distributed renewable sources are classified depending on the resource availability, load demand, and existing electrical power system, into three categories namely low penetration with existing grid, high penetration with existing grid, and high penetration with future smart-microgrid configuration.

3.2.1. Low penetration with existing grid

In low penetrated networks, the distributed generator units are not involving in frequency control activities and voltage control activities of the PCC point. Grid operator is responsible for managing the overall system stability, and DG operators can send the maximum available power to main grid and local loads without major consideration of grid constraints.

The DG operators have to deliver the power based by grid synchronization via PLL systems with correct phase sequence. Whenever grid frequency exceeds the allowable limit, the inverters are required to disconnect from the grid. And it operates in power factor (PF) correction mode, where PF keeps closer to unity. Most of PV units and wind generators can inject the maximum available active power into the grid; most existing voltage source converter (VSC) is operating in power factor correction mode (zero reactive power).

The network operators face real problem to when DG sources are connected to low-voltage lines since microgrids have dispersed generation units; sizes of the DGs are very small and low inertia characteristic, especially frequency deviations. The amount of DG units connecting to particular distribution network is limited by the voltage control margins of that distribution network; to overcome these challenges, static synchronous compensator (STATCOM), voltage source converter (VSC), automatic tap control transformers, and special control mechanism are used by operators to control the network voltage.

3.2.2. High penetration with existing grid

When growing the renewable energy source, penetration causes complication in the system constraints due to the intermittency of RES; that the percentage of the renewable power injected into the existing grid is relatively high as compared to the power assigned to the conventional power plant. Therefore, in such type of situation, intermittent power sources cannot work as passive generators, but they have to actively participate in grid frequency and voltage control activities. In addition to grid synchronization with phase sequence matching and protection system, controls and inverters should be more intelligent.

The grid operator cannot transfer the energy to or from main grid in the case of islanded power systems with a significant penetration of RES power, so the isolated system has to deal with intermittency issue. Since the amount of power delivered considerably effects the grid stability, phase-balanced operations and proper VSC inverter connection strategies have to be implemented in the system. Voltage control loop can be included in VSC inverters to provide the required reactive power to the grid, in this way VSC will intelligently response to the grid conditions. On the other hand, inverters have to operate within defined power factor range, not unit power factor so that VSC will have the capability to control the grid voltage at the PCC point.

3.2.3. High penetration with smart grid concepts

The combination of different renewable energy generation resources (such as microhydropower, photovoltaic arrays, geothermal, wind-turbine generators) in a microgrid can be integrating to the grid and increase the penetration of renewable energies to change the whole system into a smart grid with advanced technologies. Upcoming smart grid networks will provide a real-time, multidirectional flow of energy and information. Smart intelligent equipment with modern digital controls are used in entire electricity grid from central control office to end customer levels [18].

However, maintaining the stability and reliability of the network becomes a problem when the contribution from DGs is maximizing, then solution may be using smart grid concepts

http://dx.doi.org/10.5772/intechopen.78634

such as microgrids, large-scale energy storage with advanced energy management systems, smart homes with demand response management, etc. This will help in better communication and coordination between all the participants in the electricity business such as power plant operators, network operators, end consumers, and government.

3.3. Microgrid control

Generally, the control system must place at different level of the system and a consistent communication between several control units is required since there is a continuous change of power production in the DGs and the load demand in fluctuation with time. MG central controller (MGCC) installed at the medium-/low-voltage substation, which has a supervisory task of centrally control and managing the MG, integrates with the main grid. The MGCC includes several key functions, such as economically managing functions and control functionalities, is the head of the hierarchical control systems, and communicates between network operators.

The MG is intended to operate in the following two different operating conditions: the normal interconnected mode with a distribution network and the emergency mode in islanding operation via a central switch, which must also implement the synchronization between both power systems.

The typical single-line structure of a microgrid control system is described in **Figure 8**. It is clear that a direct connection of the microgrid LV line to DGRs (PV, wind generator, microturbine) and to the electrical grid network is not possible so power electronic interfaces (DC/AC or AC/DC/AC) are required due to the characteristics of the energy produced. Inverter control and circuit protection is thus an important concern in MG operation.

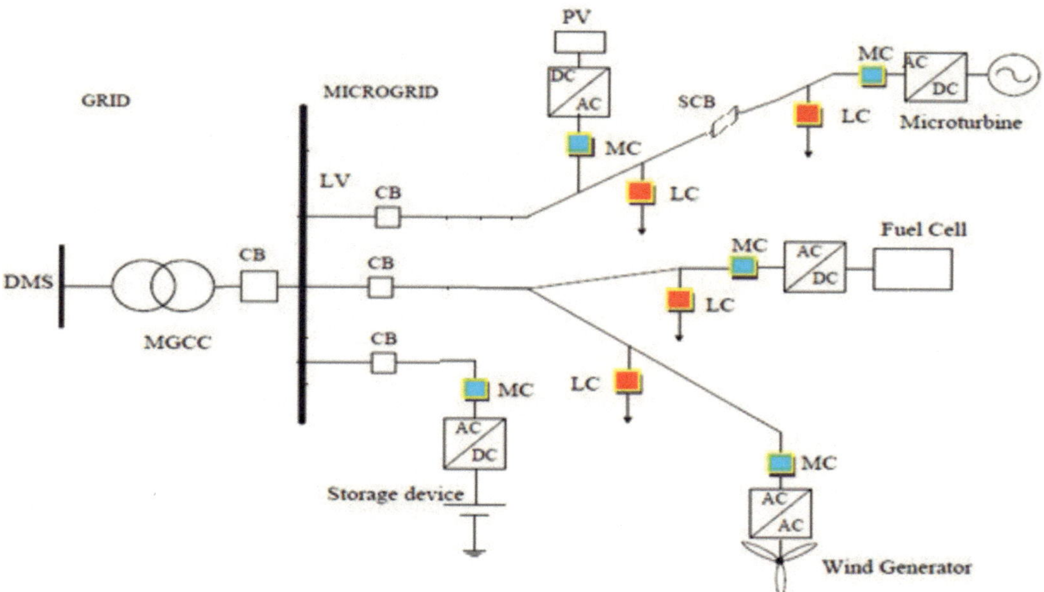

Figure 8. The microgrid control architecture [19].

In the microgrid control system, there are main parts including: microsource controllers (MCs) on the consumer production side and load controllers (LCs) on the consumer demand side; microgrid system central controller (MGCC) on the middle of the main grid; and microgrid structures and distribution management system (DMS) in the grid network side.

The different DG sources and energy storage devices are connected to the low feeder lines through the micro source controllers (MCs). MC has a function of controlling the power flow and bus voltage profile of the microsources according to the load changes or any other disturbances. These feeders are also supplied with several sectionalizing circuit breakers (SCBs) which help in isolating a part of the microgrid as needed in case trouble. Power electronics interfaces and inverters (AC/DC, AC/AC, DC/AC) are important mean for controlling and monitoring the loads using load controllers (LCs).

The overall operation and management in both the modes (isolated and grid-tied) is controlled and coordinated with the help of microsource controllers (MCs) at the local level and microgrid system central controller (MGCC) at the global level; there is a point of common coupling (PCC) through the circuit breakers (CBs) between the microgrid and the medium voltage-level utility grid. The MGCC is responsible for the overall control of microgrid operation and protection; like maintaining specified bus voltages and frequency of the entire microgrid; energy optimization for the microgrid. On the utility side, there is a distribution management systems (DMS) having several feeders including several microgrids; function for distribution area management and control.

So there are two parts in control tasks: first one is microgrid-side controller (MC and LC) to take the maximum power from the input source, and the protection of input-side converter must be considered. Second part is that grid-side controller (MGCC & DMS) which is having the following main tasks: (a) input active power control derived for network; (b) control of the reactive power transferred between network and microgrid; (c) DC link voltage control; (d) synchronization of network; and (e) assurance of power quality injected to the network [20].

3.4. Microgrid protection systems

The protection systems of a microgrid are very challenging since there is bidirectional flow of power in the system; in case of bulk power system, power flow is unidirectional. But with DG sources, the grid power flow become bidirectional; from both utility substation to microgrid energy storage and load or from local DG sources to the main grid or other microgrid, so there is a consistent reverse flow of current from maximum energy production to high energy consumption.

As a result, the old protection coordination schemes for the safety and stability in grid operations are no longer functional due to the different ways of the current flow for different operations; so innovative protection order is necessary for protecting the grid based on data collecting and information sharing to ensure proper operation in a microgrid.

The conventional overcurrent elecromechanical relay cannot sense the fault in the system since the fault current is limited and change direction depending upon the location of the

http://dx.doi.org/10.5772/intechopen.78634

fault. Sophisticated automatic protection system operation is essential for safe grid operation [20, 21]. Protection strategies can be based on communication, time grading, and other smart technologies like using microprocessors.

Differential relays offer perfect for transmission lines. They have many features over distance relays. These relays have better sensitivity.

A fast digital communication network is used to command the circuit breakers and protection relays from central control unit (MGCC). The digital current differential relays and wide area protection (WAP) are reliable and selective for the protection of microgrids when used with optical Ethernet-based communication, and wireless communications are also possible [21, 22].

The integration of distributed energy resources (DERs) into microgrid and/or into existing distribution systems faces a considerable challenge to existing power system protection. When the disturbance occurs in the utility grid system as well as in the microgrid protection system, microgrid must respond based on the location of the fault. First, when the disturbance happens in the utility grid, the protection has to trip the circuit to disconnect the microgrid from the main grid as quickly as possible by a fast semiconductor switch called static switch (SS). The other case is when the fault occurs within the microgrid, the protection system isolates the smallest possible zone of the distribution line to eliminate the fault [20].

If any events in the main grid appear, an islanded operating mode can be implemented because the electrical system is organized in the form of an MG with an MGCC. The MG islanding process may result from an intentional disconnection from the MV grid (due to maintenance needs) or from a forced disconnection (due to a fault in the MV network such as voltage dips). The disconnection is performed by a static bypass switch opening itself as a controllable load or source.

Protection system is one of the major challenges for microgrid which must react to both main grid and microgrid faults. The protection system should cut off the microgrid from the main grid as rapidly as necessary to protect the microgrid loads for the first case, and for the second case, the protection system should isolate the smallest part of the microgrid when clears the fault [23, 24].

4. Conclusion

The integration of microgrids with RES in the current utility grids is the first step toward the transition from the conventional power system to smart grid system. The main barrier to expand this technology is lack of readiness to change our dependence on the coal and oil-driven economy and life style, but the technology has been advancing, the Internet of things makes our societies to decide on different life choice. Most of the existing power system overall cost is also becoming expensive in the near future; RES technological improvement; advancement in energy storage systems can help the new microgrid system based on DG to become economically viable to consumer. More penetration of RESs is expected in microgrid systems as they are almost pollution-free and thus environment friendly.

Microgrid is well known in North America and Europe and used in those developed countries; however, there will be positive progress in less developing country to build their electricity and power infrastructure in a futuristic microgrid and smart grid model, one know that there is big financial obstacle and skill gap in those country but if there is the willingness from the government to transform their development plan into small-scale microgrid in power and heat demand.

In this chapter, the overall structural components of hybrid power system and the major challenges in the integration of microgrids like control and protection microgrids are presented. A significant research and development is required to transform and implement a microgrid system.

Acknowledgements

The authors are obliged to the authorities of Institute of Technology (IoT), Hawassa University for providing the enabling environment and support to carry out this work.

Conflict of interest

The authors declare no conflict of interest.

Notes/thanks/other declarations

The chapter is a collaborative effort of the authors. The authors have contributed collectively to the theoretical analysis, literature review, and manuscript preparation.

Author details

Mulualem T. Yeshalem and Baseem Khan*

*Address all correspondence to: baseem.khan04@gmail.com

School of Electrical and Computer Engineering, Hawassa University Institute of Technology, Hawassa, Ethiopia

References

[1] Khan B, Singh P. Selecting a meta-heuristic technique for smart micro-grid optimization problem: A comprehensive analysis. IEEE Access. 2017;**5**:13951-13977

[2] Rezaei MM, Soltani J. A robust control strategy for a grid-connected multi-bus microgrid under unbalanced load conditions. Electrical Power and Energy Systems. 2015;**71**:68-76

[3] Khamis A, Shareef MH, Ayob A. Modeling and simulation of a microgrid testbed using photovoltaic and battery based power generation. Journal of Asian Scientific Research. 2011;**2**:658-666

[4] Fitgerald AE, Kingsley JC, Umans SD. Electric Machinery. New York: McGraw-Hill; 1990

[5] Momoh J. Smart Grid: Fundamentals of Design and Analysis, Institute of Electrical and Electronics Engineers. 1st ed. New Jersey, United States: John Wiley & Sons, Inc; 2012

[6] Jariso M, Khan B, Tesfaye D, Singh J. Modelling and designing of stand-alone photovoltaic system (case study: Addis Boder Health Center south west Ethiopia). In: IEEE International conference on Electronics, communication, and aerospace Technology, ICECA 2017; Coimbatore, India; 20-21 April. 2017. pp. 168-173

[7] Phrakonkham S, Le Chenadec J-Y, Diallo D, Remy G, Marchand C. Reviews on microgrid configuration and dedicated hybrid system optimization software tools: Application to Laos. Engineering Journal. 2010;**14**(3):15-34

[8] Yeshalem MT, Khan B. Design of an off-grid hybrid PV/wind power system for remote mobile base station: A case study. AIMS Energy. 2017;**5**(1):96-112

[9] Omari O, Ortjohann E, Mohd A, Morton D. An online control strategy for DC coupled hybrid power systems. IEEE Power Engineering Society General Meeting. 2007:1-8. ISSN: 1932-5517

[10] SIEMENS USA. Smart Grid Solution. Available from: https://w3.usa.siemens.com/smartgrid/us/en/microgrid/pages/microgrids.aspx [Accessed: April 5, 2018]

[11] Singh P, Khan B. Smart microgrid energy management using a novel artificial shark optimization. Complexity. 2017;**2017**:Article ID 2158926: 22 p

[12] Setiawan AA, Zhao Y, Susanto-Lee R, CVN. Design economic analysis and environmental considerations of mini-grid hybrid power system with reverse osmosis desalination plant for remote areas. Renewable Energy. 2009;**34**(2):374-383

[13] Edward JN, Ramadan A, El-Shatshat. Multi-Microgrid Control Systems (MMCS). IEEE Power and Energy Society General Meeting. 2010;**25**:1-6

[14] Saleh MS, Althaibani A, Esa Y, Mhandi Y, Mohamed AA. Impact of clustering micro grids on their stability and resilience during blackouts. In: 2015 International Conference on Smart Grid and Clean Energy Technologies (ICSGCE); 2015. pp. 195-200

[15] El-Khattam W, Salama MMA. Distributed generation technologies, definitions and benefits. Electric Power Systems Research. 2004;**71**:119-128

[16] Khan B, Singh P. Optimal power flow techniques under characterization of conventional and renewable energy sources: A comprehensive analysis. Journal of Engineering. 2017;**2017**:16. Article ID: 9539506

[17] Kifle Y, Khan B, Singh J. Designing and modeling grid connected photovoltaic system: (case study: EEU building at Hawassa city). International Journal of Convergence Computing. 2018;**3**(1):20-34

[18] Huang JY, Jiang CW, Xu R. A review on distributed energy resources and microgrid. Renewable and Sustainable Energy Reviews. 2008;**12**(9):2472-2483

[19] Passey R, Spooner T, MacGill I, Watt M, Syngellakis K. The potential impacts of grid-connected distributed generation and how to address them: A review of technical and non-technical factors. Energy Policy. 2011;**39**:6280-6290

[20] Mohamed Y, El-Saadany EF. Adaptive decentralized droop controller to preserve power sharing stability of paralleled inverters in distributed generation microgrids. IEEE Transactions on Power Electronics. 2008;**23**:2806-2816

[21] Mumtaz F, Bayram IS. Planning, operation, and protection of microgrids: An overview. In: 3rd International Conference on Energy and Environment Research, ICEER 2016, 7-11 September 2016; Barcelona, Spain: Elsevier Energy procedia; 2017. pp. 94-100

[22] Eissa MM. New protection principle for smart grid with renewable energy sources integration using WiMAX centralized scheduling technology. Electrical Power and Energy Systems, Elsevier. 2018;**97**:372-384

[23] Guerrero JM, Vasquez JC, Matas J, Castilla M, de Vicuna LG. Control strategy for flexible microgrid based on parallel line-interactive UPS systems. Industrial Electronics, IEEE Transactions. 2009;**56**(3):726-736

[24] Odavic M, Sumner M, Zanchetta P, Clare JC. A theoretical analysis of the harmonic content of PWM waveforms for multiple-frequency modulators. IEEE Transactions on Power Electronics. 2010;**25**(1):131-141

Chapter 5

Architecture Parallel for the Renewable Energy System

Nguyen The Vinh, Vo Thanh Vinh, Pham Manh Hai,
Le Thanh Doanh and Pham Van Duy

Additional information is available at the end of the chapter

http://dx.doi.org/10.5772/intechopen.78346

Abstract

This chapter present one possible evolution is the parallel topology on the high-voltage bus for the renewable energy system. The system is not connected to a chain of photovoltaic (PV) modules and the different sources renewable. This evolution retains all the advantages of this system, while increasing the level of discretization of the Maximum Power Point Tracker (MPPT). So it is no longer a chain of PV modules that works at its MPPT but each PV module. In addition, this greater discretization allows a finer control and monitoring of operation and a faster detection of defects. The main interest of parallel step-up voltage systems, in this case, lies in the fact that the use of relatively high DC voltages is possible in these architectures distributed.

Keywords: renewable energy systems, photovoltaic system, DC-DC step-up converter, distributed generator, power line communication

1. Introduction

In a constant effort to improve systems, numerous studies have used to find the best configuration to make photovoltaic systems more performing in terms of robustness and reliability. It now appears that systems distributed intelligence associated with distributed converters, constitute one of the most promising solutions [1–3]. Also our study will focus on the facilities of small powers that have good potential in terms of evolution. For those whom it concerns the higher power installations, the problems arise in another way. In fact, maintenance and production follow-up are provided by specialized teams, which is not always the case for small installations. We have also mentioned the possibility of opting for DC bus in the

modularity of power generation. It is therefore more simply to consider transpose to small installations the modularity that is required in large power plants photovoltaic production.

One possible evolution is the parallel topology on the high-voltage bus shown in **Figure 1**. The system is no longer connected to a chain of PV modules but directly to the output of the PV module. This evolution retains all the advantages of this system, while increasing the level of discretization of the maximum power point tracking (MPPT). So it is no longer a chain of PV modules that works at its MPPT but each PV module. In addition, this larger discretization allows finer control of operation and faster fault detection.

The main advantage of parallel systems to step-up in the case that concerns us here is the fact that the use of relatively high DC voltages is possible in these distributed architectures, as also mentioned in their theses Estibal [2], Cabal [4], and Vighetti [5, 6]. The distributed structure is very advantageous both from the point of view of optimization and robustness to defects. It is also a modular application that allows for the multiplication and diversification of technologies, for example the combination of several types of photovoltaic sensors different with different renewable sources such as wind, hydraulic, compressed air, biomass.

Figure 1 shows a possible configurational aspect. This is a DC voltage bus that can reach 1 kV. All panel-converter elements are connected in parallel "dubbing" on this bus. The

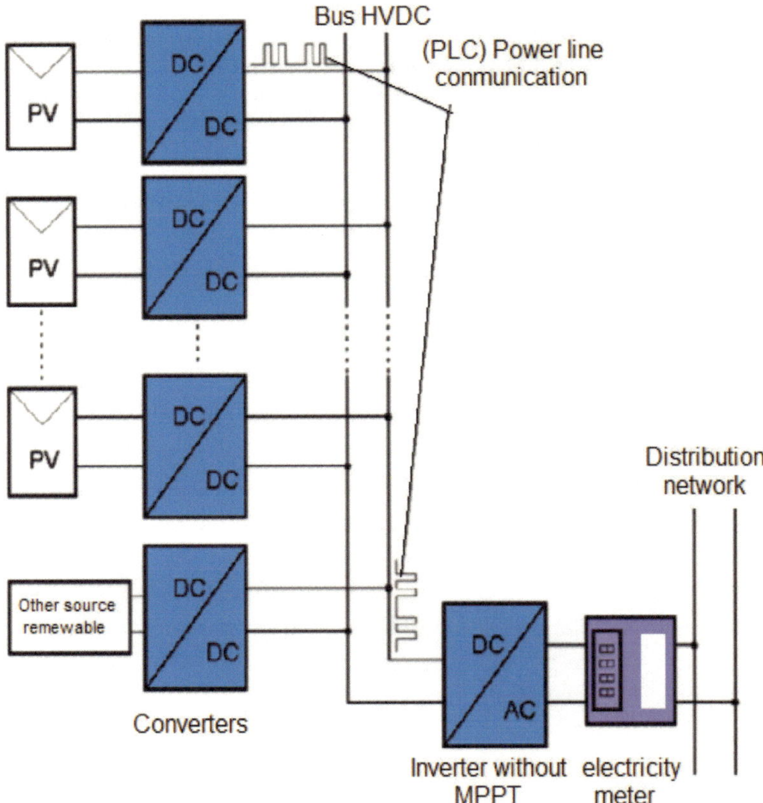

Figure 1. Diagram of a parallel structure on a continuous high-voltage bus.

phenomena of susceptibility to electromagnetic impulses (IEM) of lightning strikes on the DC bus are minimized by the use of twisted or very close cables, presenting only very small surfaces exposed to magnetic fields. Note the need for proper monitoring to avoid islanding. The downstream inverter does not manage the control of the overall MPP, as each panel is managed locally optimally. The case where several inverters are connected in parallel, for a question of power, does not change the question of control. The use of a high voltage makes it possible to envisage a reduction in the section of the cables, which constitutes a material gain in copper (or aluminum if necessary). The high voltage principle can be applied for voltages up to more than 1 kV. Some studies plan to push the limits to 8 kV [7], especially for the transport of energy produced by wind turbines. Raising the tension of a panel has many advantages. Without wishing to list them all, and at the risk of quoting the announcements of manufacturers present in this niche, we will retain some of the most remarkable aspects, namely:

- The constant output voltage of the converters makes it possible to directly attack an inverter.
- The resistances of the connectors are less critical at low currents.
- Contact quality problems are minimized by the use of high voltage. It is possible to use smaller cable sections as voltage is increased, resulting in a saving of copper (or aluminum).
- It is possible to use a DC voltage at the output of the converter.
- DC-DC risers are remotely controllable and are compatible with power line communication (PLC).
- The converters can locally provide the MPPT, which eliminates this feature of the inverter.

On the other hand a certain number of problems are inherent to the high tensions:

- Continuous high voltages present significant risks in terms of fire, which imposes a suitable security device.
- The insulation must be neat.
- A single inverter is inefficient when the powers to be converted are low.

This system using a high-voltage DC bus will allow, with appropriate adaptation, the supply of energy from several sources such as photovoltaic, wind, etc. to the output inverter. Although this technique was described many years ago, the generation of high DC voltages with good efficiency is not easy especially for DC-DC inverters with an output voltage greater than 10 times the voltage input.

1.1. Serial converter connection

The galvanic isolation of the output allows several DC-DC converters to be connected in series by simply connecting the positive output of one converter to the negative input of the other (**Figure 2**). In this way, non-standard voltage rails can be generated, however, the output current of the high voltage converter should not be exceeded. If the required output voltage of a converter is greater than the nominal voltage of the next converter, the outputs of both

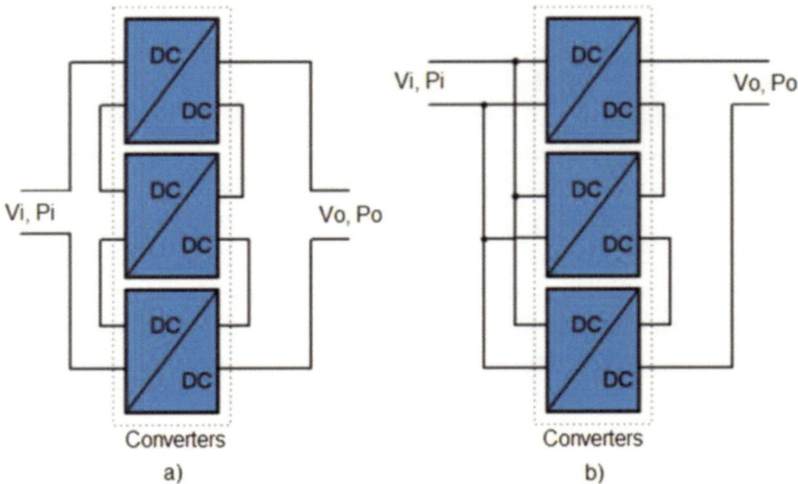

Figure 2. Serial connection of converters: (a) serial inputs and outputs, (b) parallel inputs and serial outputs.

converters can be connected in series in order to reach the desired output voltage. It is recommended that two converters of the same model be used for this application. The high isolation of the converter outputs allows multiple converters to be connected in series by simply connecting the positive output of one converter to the negative terminal of the other as shown in **Figure 2** [8, 9]. For this series mounting, the designer must also ensure that the total output voltage does not exceed the output breakdown voltage of the output for each converter.

When the converters are connected in series, additional external filtering is strongly recommended because the converter switching circuits are not synchronized. It is possible that there is a phase summation of the ripple voltages of the two converters resulting in relatively high beat frequencies.

1.2. Connecting converters in parallel

In power augmentation applications, DC-DC converters are often connected in parallel (**Figure 3**) to form a more robust power distribution system and produce higher currents. This distribution mode is commonly used for applications requiring high currents. The voltages can also be very low, for example in the case of a microprocessor supply. Compared with a single high-power converter, paralleling allows, by a homogeneous distribution of power, to reduce the stress endured by semiconductors and thus to improve reliability, robustness and service life conversion stage. This structure also provides a degree of freedom, in terms of flexibility and modularity, compared to a conventional converter.

The energy transfer converters cannot have exactly the same electrical characteristics, because of a dispersion of characteristics on the electronic components constituting them and a slight difference of connections. In operation, this causes a natural imbalance on the distribution of currents between each converter. Thus, the probability that one or more converters operate with an excess of current compared to others is great. This phenomenon results in significant thermal stress in the most stressed semiconductors, so the robustness, reliability and service life of the system are reduced, canceling the initial advantages of the structure. In order to remedy this problem and to guarantee a homogeneous distribution of the current and therefore of the power, current regulation is indispensable in these parallel structures [10].

1.3. Cascading converter connection

Theoretically, the voltage gains of an inductively accumulating DC-DC converter tend to infinity for a unit duty cycle. In reality, this gain is limited by the series resistances of the components (inductances and capacitors) [11, 12]. This gain cannot, or difficulty, exceed 5 for a high duty cycle. The specifications impose a voltage gain of around 25. A simple DC-DC boost converter cannot achieve this gain. To obtain this voltage gain, it is necessary to use a cascading arrangement (**Figure 4**).

After introducing renewable energy resources for conversion into electricity, we have examined in this chapter the parallel structures of high voltage DC voltage booster converters for the optimization of energy transfer management renewable sources by power line communication (PLC). The different structures were presented considering the possibility of improving the efficiency of the system. This system has the advantage of communication between smart converters in the energy transfer process. The study of the modalities of this communication is the subject of this chapter.

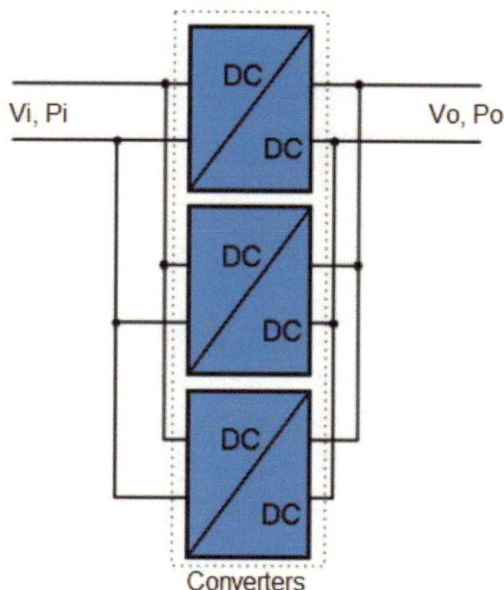

Figure 3. Connecting converters in parallel.

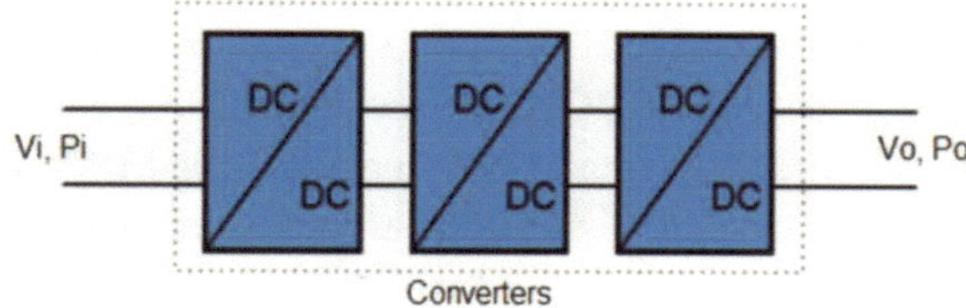

Figure 4. Connection of converters in cascade.

2. Distributed architecture for energy management and supervision of the HVDC bus by PLC

Investing and exploiting electricity distribution networks generated from renewable energy sources such as solar, wind, ocean waves, biomass ... is a technological development to improve efficiency and stability. of all switching systems and power supply. With this affordable power distribution architecture, Smart Grid integration is a new option to optimize the output of each power source. This approach requires the addition in the system of intra and inter-communication tools with a central controller. A technical solution is power line communication (PLC) using the existing power line infrastructure for communication purposes. This new management capability improves performance and stability throughout the energy transfer process. In recent years, new energy supply concepts have been introduced. One of them newly introduced in renewable energy systems are smart grid concepts [13–15]. In addition, nowadays power-engineering field is facing huge challenges since the growing interest in intermittent renewable energies impose a whole new paradigm of operation. The use of these resources must be properly implemented to ensure safe, effective, autonomous and sustainable operations. At the same time, distributed energy sources are connected to each other to form a smart grid to provide adequate and reliable energy. Distance determination is made to provide energy for electricity used. Many of the energy sources in this proposal are of a continuous type and can be transformed before performing DC-AC transformations. With strong incentives for green energy, availability and advances in storage technology, the power generation in this smart grid proposal applies not only to distribution networks with coal or gas and burning, but also an incentive for anyone who wants to produce their own electricity. Data transmission is also required in the proposed Smart Grid concept based on the high voltage DC distribution system HVDC from renewable sources. The data transmission is necessary for the necessary functionalities and applications that will be integrated in the system, for example for the monitoring of the parallel DC-DC converter with power failure management and protection. For this reason, PLC is a possible alternative.

A possible answer to these challenges may be the solution proposed in **Figure 1**. In this figure presented in the above section, a power system consisting of various sources such as solar cells with different technologies, wind turbines and the system or any other dc power source interconnected by a high-voltage direct current DC storage bus compressed air energy (CAES); A regional power plant is composed of many small generators for internal lines, and a low cost overall structure [16, 17]. This structure shows the power on the HVDC line emitted from high-performance DC-DC converters. In the AC distribution network, a large DC-AC converter communicates between the HVDC and the AC network. Therefore, the HVDC system is designed to replace the traditional medium-voltage (MV) branch of the overhead line and distribute low-voltage AC by the HVDC distribution system. In this distribution network, the MPPT should be integrated in each DC-DC converter. In addition, these MPPTs optimize performance from each source, and the overall performance of the MPPT increases with respect to the central structure. In industry, the distribution and production systems of DC and HVDC are widely used and developed to replace AC power to DC. Moreover, there is an advantage of DC voltages for smart grids that do not need to be synchronized for decentralization.

In AC systems combined with intermediate HVDC lines, the quality of alternating current decreases at the output of the transmitter. This can be improved by reducing the number of branches of the MV grid and reducing the length of the line. Overhead lines can be replaced by underground low voltage cables. This has directly led to a decrease in the amount of defects generally occurring in MV networks. As a result, the quality and reliability of the distribution grid is improved and economic losses due to increased power quality are reduced. The HVDC bus with the configuration of the communication system is proposed when communication is made with the PLC. Therefore, the structure and configuration of the HVDC grid is related to a specific physical geographic area. This will bring restrictions to the controller, such as the reliability of the communication chain, the PLC loop is inserted. In addition, the inverters in the bus generate harmonics and interference on the transmission channel. All of these, along with some of the features that are implemented in the content grid to determine the boundary conditions and minimum requirements for PLCs. Thus, we can summarize that the main objectives of the development of distribution systems and continue for smart grids are the profitability and reliability of electricity distribution. Ubiquitous communication plays a key role in smart grids and HVDC concept presents a new approach for smart grid implementation [18].

The parallel modular structure topology shown in **Figure 5**. This structure is no longer connected to a chain of PV modules but directly to the output of the PV module. This evolution retains all the advantages of the "row" structure, while increasing the level of discretization of the MPPT. So it is no longer a chain of PV modules that works at its MPPT but each PV module. A productivity gain is therefore to be expected with respect to the "row" structure. In addition, this larger discretization enables finer monitoring and faster fault detection.

With medium and larger systems, it may indeed be advantageous to carry out MPPT initially by a DC-DC converter for each PV chain. Their outputs are connected to a DC bus, from which the synchronous inversion gate is made by a central inverter. If the nominal voltage of this DC bus is standardized, there is the possibility of integrating other generator sets (such as modular hydraulic systems or small wind turbines) as well. **Figure 6** illustrates the general concept of such a system.

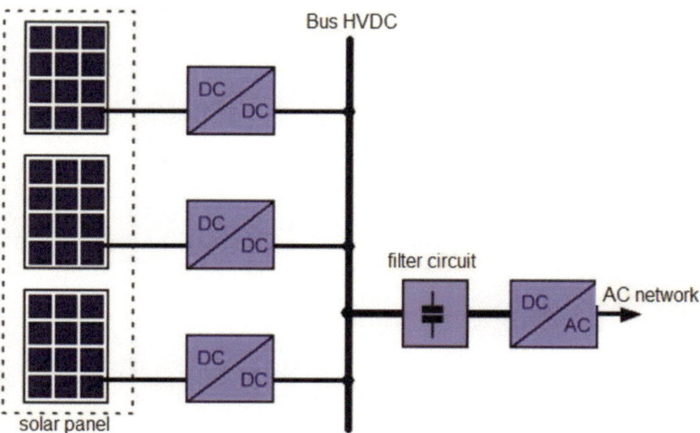

Figure 5. The main diagram circuit of a photovoltaic plane operating according to the mentioned method.

As shown in **Figure 6**, there are normally several DC-DC converters connected to the DC bus on the side of the generator, while the grid side of a central inverter goes on to power on the grid of electricity. The inverter should take exactly the amount of power from the DC bus that is fully supplied to the side of the generator.

Different operating states occur when the system starts up or when more power is available on the side of the inverter generator is capable of processing. How are these operating conditions managed? The requirements of stability, forgivingness, independent manufacturer compatibility and simple system design, can hardly be met by a digital bus system and a central control unit. The different operating conditions are rather to be detected by each component involved in an autonomous way. The necessary information will be extracted from the evaluation of the voltage level of the actual DC bus. **Figure 6** shows the operation of the ranges defined for this purpose.

The main hard point of this structure is the great ratio of elevation between the output voltage of the PV module types, different generators and the voltage required for the injection on the distribution HVDC bus. Indeed, for a non-isolated DC-DC converter the higher the elevation ratio the greater the losses. When this ratio is too important (>8 in general), it is necessary to use isolated structures or cascades of converters. In this case, the necessary elevation ratio is close to 10, which limits the efficiency of the DC-DC converter and penalizes this topology.

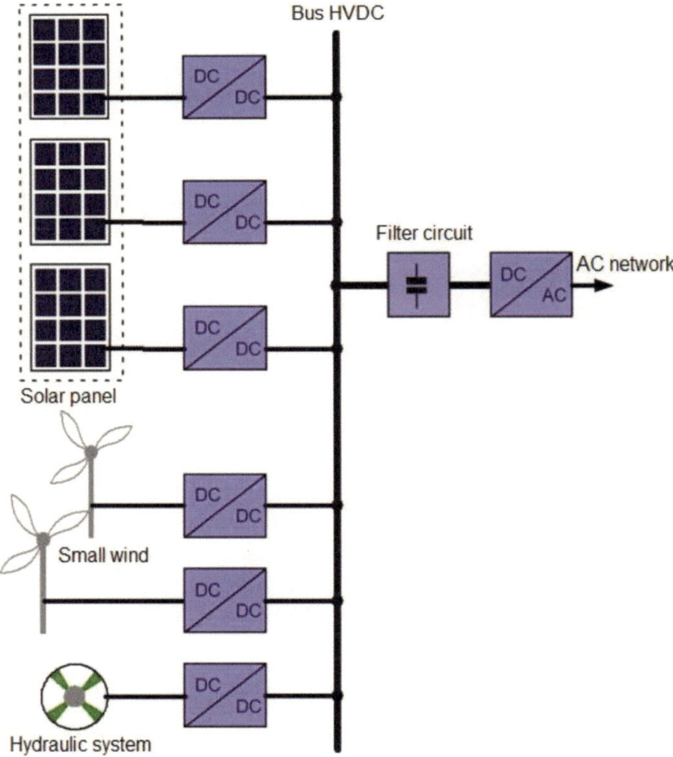

Figure 6. The general concept of a distributed electric power generation system.

One of the main advantages of the HVDC system is the rapid controllability of the transmitted power. The power control in the HVDC system is based on current or voltage control, while minimizing losses in the DC line, it is important to maintain a constant voltage and establish the desired current on the DC line. The line from the rectifier to the inverter in order to obtain the desired power flowing in the DC line by varying the voltage at the converters. This method is important for voltage regulation in HVDC system to meet the examination of the optimal use of insulation and as we have noticed that the voltage drop in the DC line is reduced compared to AC the line due to the lack of reactive voltage drop.

Monitoring the transmission of energy with MPPT from separate renewable sources reliable, safer, more efficient and save. However, the installed consumers, the peak electricity demand would be so great that beyond the 3 kW power of an electrical system. The practical limit for the renewable energy transmission system is high that its output voltage to the inverter is generally low, beyond which the diameter of the wire becomes too large and cumbersome to handle. For a 60 V system, 50 amps implies power consumption almost 3 kW, it is obvious that a higher source voltage is needed to meet future demand on 3 kW. Without exceeding the limit of 50 amperes, 400 V is used. 400 V has been internationally agreed upon the maximum voltage that can be used safely in a conventional vehicle system without additional protection and also a 400 V system is capable of providing up to 19 kW without exceeding the current limit of 50 A of the wires. The general structure of a 400VDC power system is shown in **Figure 7**.

We present simple hardware of distributed architecture based on technical and performance requirements, proposed PLC architecture for HVDC currents. Commercial LC techniques and characteristic signaling methods are used. HVDC network links are recommended and evaluated by meeting the requirements set for that application. The distance between the mesh elements is evaluated by theoretical analysis and the actual data transfer test. Based on the DC-DC optimizer incorporates interface transformations, a PLC-based network architecture for HVDC line systems to meet all of these requirements. Note that a self-test step can also be added to the optimizer, but only the new PLC enhancements described in this section, the power conversion features, with their MPPT algorithms.

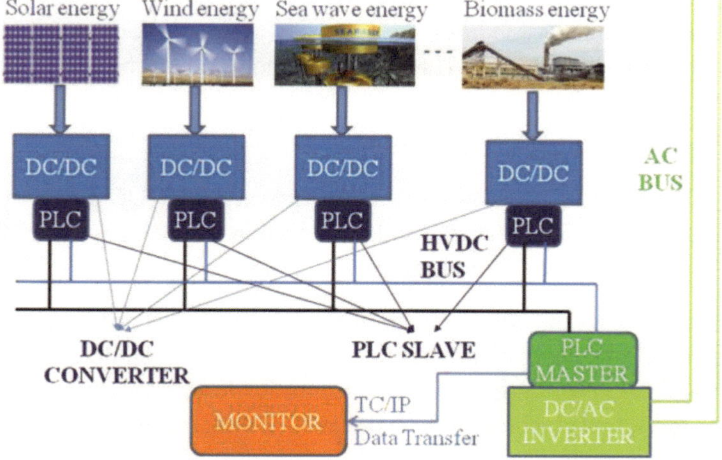

Figure 7. Offer PLC system on the HVDC bus.

PLC systems are installed in HVDC lines using clean inter-wafer circuits, used for insulation and impedance between the DC-DC converter and the power grid. It is possible to view PLC systems as an additional part of each converter, without modifying its basic structure. Nevertheless, the two main steps are driven by a common microcontroller peripheral interface (PIC) controller assuming both the controller's master–slave tracking and control functions. This system is required for communication in small and medium power systems, such as remote reading, fire/fire alarm control and shutdown. The system is designed with a digital modulator [19] to reduce the workload of the main controller of the DC-DC-PLC converter. These are two independent steps and a single control output reduces the cost of the entire system significantly.

3. PLC system design on the HVDC bus

Renewable system manufacturers' energy products increasingly differentiate their products by providing more sophisticated and inventive features such as safety, stability, control, comfort, convenience and performance. However, the use of these applications requires high volume data exchange and a reliable data communication network to enable efficient and efficient control over electronic devices. On the other hand, conventional infrastructure systems and distributed decentralized power generation systems transmit on high voltage lines, the PLC system does not affect the quality of energy transmission of the system. In addition, the nodes of the PLC slaves will be integrated in the DC-DC converter corresponding to each generator. Therefore, the design of the PLC system on the high voltage line is the simple, economical and reliable signal transmission.

In order to simplify and standardize the electrical system of the renewable energy generators on the HVDC bus, a number of communication standards and protocols have been proposed. Most communication systems on the HVDC bus are modified based on existing commercial communication technologies to meet the HVDC bus automation requirements such as high stability and error correction capability. In general, almost all networks for the HVDC bus are digital networks because of its high stability in terms of noise protection and error correction capability. From distributed architecture to decentralized renewable energy sources on high-voltage direct current transmission lines, we proposed to use Modbus protocol to implement the PLC system that we are designing. The Modbus protocol will be benthic in the next section.

3.1. Proposed DC power bus communication system

The basic idea of a power line communication system on the HVDC bus is to establish data communication on the DC line without installing additional wires. The simplified circuit of a PLC line carrier network is shown in **Figure 8**.

Figure 8 demonstrates a simplified conventional circuit of injection of a sinusoidal wave or rectangle carrier signal to a DC power line with separation. Node 1 and the slave node and the master node are both in-line carrier communication transceivers which consist of a power amplifier for signal transmission on the HVDC bus. In this figure, V_{conver} is the voltage of the DC-DC converter, R_{conver} is the output resistance of the DC-DC converter, V_{hvdc} is the voltage of the DC high voltage line, V_{line} is the inverter input voltage, R_{line} is the input resistance of the inverter, the signals on the line are the sine or rectangle to be injected into the power supply line and is amplified by a power amplifier. The transceiver output stage consists of an amplifier and a PLC interface signal circuit. DC carrier communication is a communication technology that makes use of the internal resistance of the power source and the parasitic components of the power cables and loads. The internal resistance of the power source is a designation factor of the performance of the communication from the power source is usually the component connected to the power line to the lowest impedance, the bulk of the power of the emitted signal would be dissipated to the power source. When node 1 is about to transmit a signal to the master node through the DC power line, the transited signal would first be amplified by the amplifier. The amplifier is usually a power amplifier to provide the signal power is appropriate because of the low impedance characteristic of the DC power line.

The transmitted signal moves through the cables of the DC line and is attenuated along the cables, a large percentage of the signal power is fresh at the impedance and parasitic components of the power line. By current the power line, the transmitted signal is attenuated and is

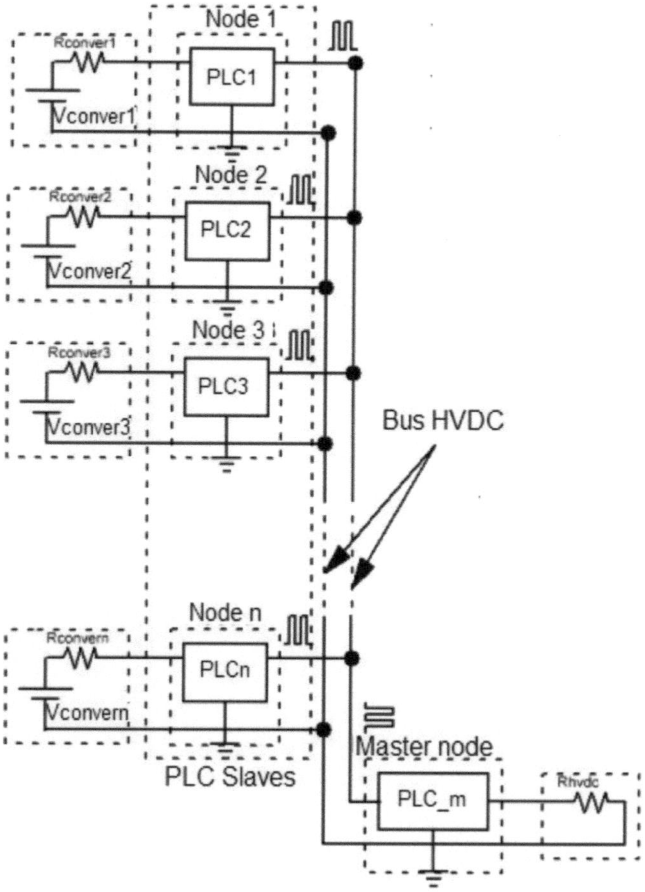

Figure 8. The general configuration of the renewable energy generator of transmission by DC-DC converters on the HVDC bus communication system with communication nodes.

sometimes distorted depending on the conditions of cable length and load of the power system. Upon reception, the detector reconstructs the deformed signal and filters the noise. The signal transmitted by the PLC slave node (1 − n) via the DC power line is then regenerated. **Figure 9** shows the simplified master–slave circuit PLC for using an amplifier to transmit the carrier signal to an HVDC bus.

3.2. Simple line current (PLC) on the HVDC bus

The basic operating principle of the proposed bearer transmitter was discussed. To facilitate "real" communication of dc line carrier data, other than the carrier signal transmitter, the system should consist of a dc line, at least one signal transmitter and a receiver signal, which is a simplex communication system. The simplified system design of DC simplex carrier communication using the proposed in line carrier transmitter is illustrated in **Figure 10**.

The function of the circuit shown in **Figure 9** is to transmit a data stream from the transmitter side to the receiver side, which is a very simple communication system that the data flows only in one direction, starting from the transmitter to the receiver. On the data stream is sent to the transmitter, it is coded, amplitude modulated and finally transmitted to the power supply network using the proposed carrier transmitter. Traveling through the power wire, the carrier signal is decoupled via a decoupling circuit at the receiver and demodulated in the original data stream. A simplex data transmission is then complete. The rate of the data system is set using 1 kbps ASK (amplitude shift keying) modulation [20]. ASK's help in the system is due to its simple modulation and demodulation processes. We have proposed a simple PLC, that the sender and the receiver are the same interface circuit so that other than the resistance and

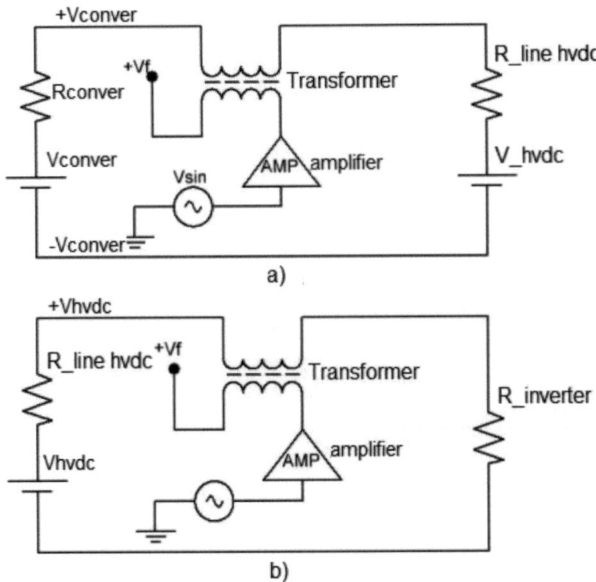

Figure 9. The carrier signal transmission circuit to an HVDC bus using an amplifier: (a) simplified PLC slave circuit; (b) simplified CPL master circuit.

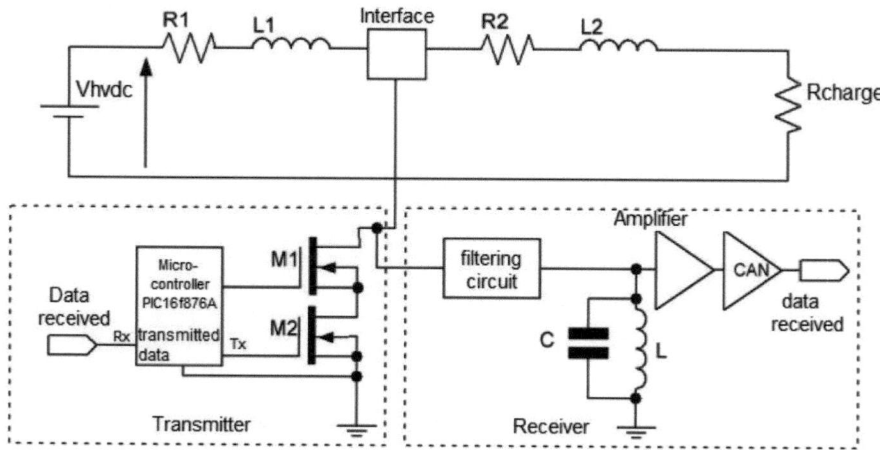

Figure 10. Diagram of the DC carrier communication system proposal with simple communication.

parasitic components of HVDC bus cables are not influencing that can be neglected. The resistors R1, R2 and Rcharge are three purely resistive loads connected to the HVDC bus.

To facilitate communication on the HVDC bus, the internal resistance of the DC power source (in this case the DC-DC converter) must exist from the converter is normally the component that carries the lowest impedance connected to the power line, most of the signal power delivered by the transmitter would be dissipated by the source into the DC-DC converter rather than distributed to receivers on the power line. This problem is important in communications on the DC system with a heavy load as the internal resistance is usually low in high voltage converter. To solve the problem, an inductance (this inductance Ls seen in **Figure 11**) can be added between the converter and the HVDC bus to increase the efficiency of the impedance of the DC-DC converter; therefore, the decrease in signal power dissipates to the converter. The value of the inductance is necessarily to optimize the value, a core inductor with a few tens of mH inductance is sufficient to increase the impedance of the converter.

The simulation circuit is shown in **Figure 12**. It is a DC simplex carrier communication system with a signal transmitter and a receiver. The system is powered by a 400 V converter, which is made up of 400 V V_{hvdc} voltage source with a very low internal resistance so the jump. Since the resistor and parasitic inductance exist in virtually the cables and the transmitter and the signal receiver are always set apart along the power cable R2 are added to simulate the basic characteristic components of the cables. Since in practical power systems, the state of charge can be very complicated and unpredictable, to make the simulation simple, it is assumed that the resistor R_{charge} is the only load connected to the net of power with the filter capacitor C1 connected in parallel. The operating principles and design considerations for a 400VDC net power communication system with multiple loads would be discussed in the last sections.

The simulation results are obtained using Pspice, which presents accurate models of the two passive and active components such as diodes and MOSFET transistors. The simulation describes how a digital signal flow is modulated in amplitude and transmitted on DC net power using the transmitter and the proposed support receiver in the steady state. Carrier

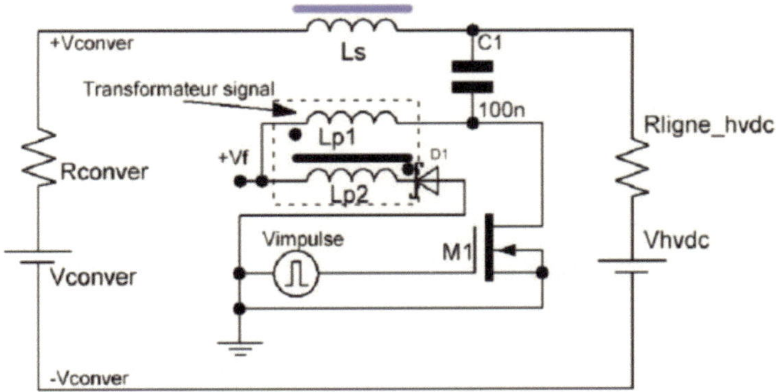

Figure 11. Schematic of the interface solutions in the PLC transmitting stage.

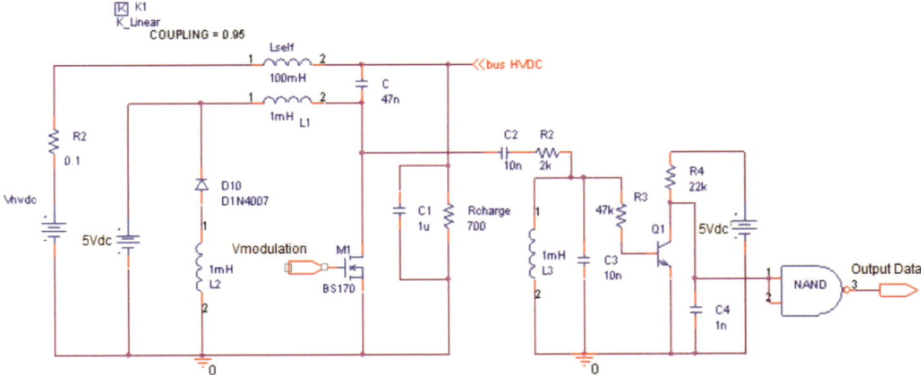

Figure 12. Simulation circuit of a simple communication system on the 400 V bus.

signal detection, demodulation, and error handling are not included in the simulation because these operations are performed by the program stored in the microcontroller. **Figure 13(a)** illustrates the waveform of the original data stream and the corresponding waveform of the amplitude modulated signal on the HVDC bus. In the simulation, the original data stream is implemented using a programmable logic pulse source "0" and "1" logic are represented by 0 and 5 V respectively. Since the transmitter circuit is only active when data transmission is required, in the simulation, the transmitter is off when the input signal is at zero volts. **Figure 13(b)** shows the ripple voltage of the HVDC bus during the transition to change the amplitude modulation. The transition time from logical state "1" to logical "0" is 10 μs. **Figure 13(c)** and **(d)** represents the voltage waveform of the inductor L2 and the drain of the MOSFET transistor M1 is compared to the original digital signal modulation.

The voltage regenerated by the current sensing circuit as a function of the filter capacitance current is shown in **Figure 14(a)**. **Figure 14(d)** shows the final output voltage of the entire receiver circuit. The output of the reception circuit is a series of rectangular pulses, where the time interval is equal to the modulation amplitude of the carrier signal; it is not yet demodulated since the signal demodulation and decoding tasks are performed by the microcontroller.

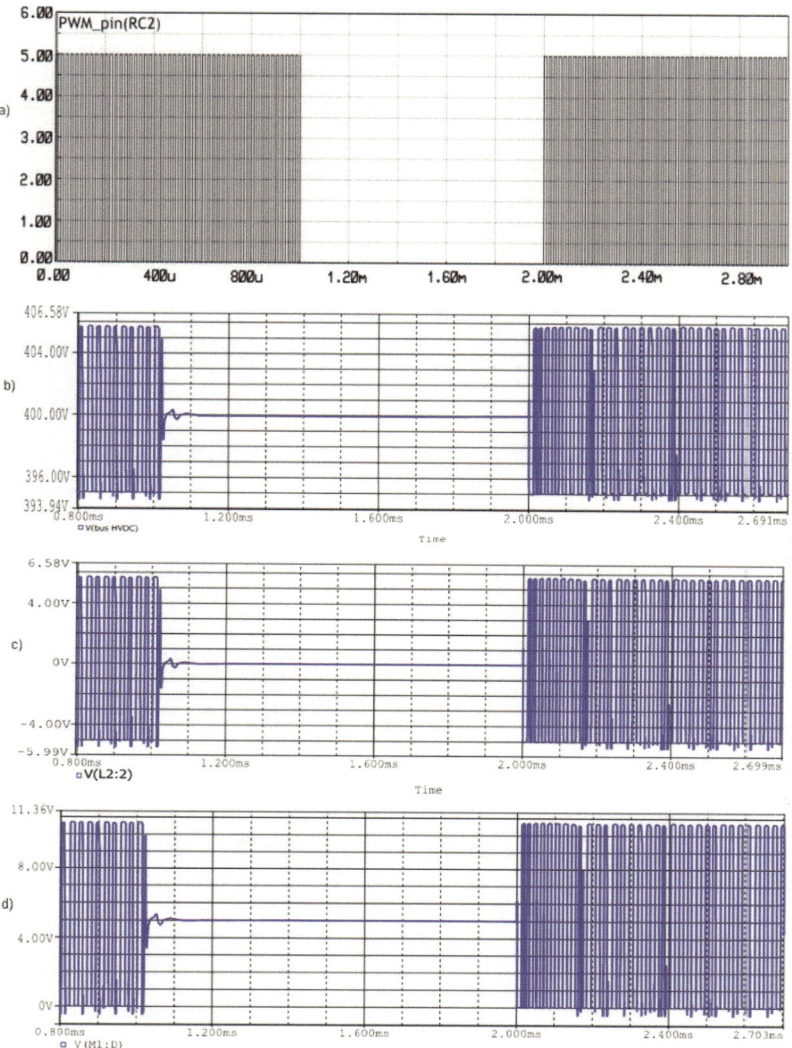

Figure 13. Simulation results at the transmitter: (a) PIC 16f876A microcontroller modulated signal, (b) 400V HVDC bus signal, (c) Signal curve at transformer inductance Lp2 also katot diode, (d) Amplified signal at drain MOSFET.

According to **Figures 13(d)** and **14(d)** in the figure, the control signal of the MOSFET transistor M1 is compared to the output signal of the receiver circuit. It is observed that the original data stream is successfully modulated, transmitted through 400VDC net power using the proposed transmitter and retrieved using the proposed receiver.

In the simulation, the signal modulator consists of a similar behavior model, the gain stages and limiters are implemented by ideal components. Since these elements practically do not exist, to verify the performance of the proposed net power DC communication system, an experimental device was built with the parameters indicated: the carrier frequency 50 kHz, the amplitude modulation at "0" equal to 0 V amplitude, the amplitude modulation at "1" equals the amplitude ±5 V, and the resistance of the HVDC bus equal to 10.

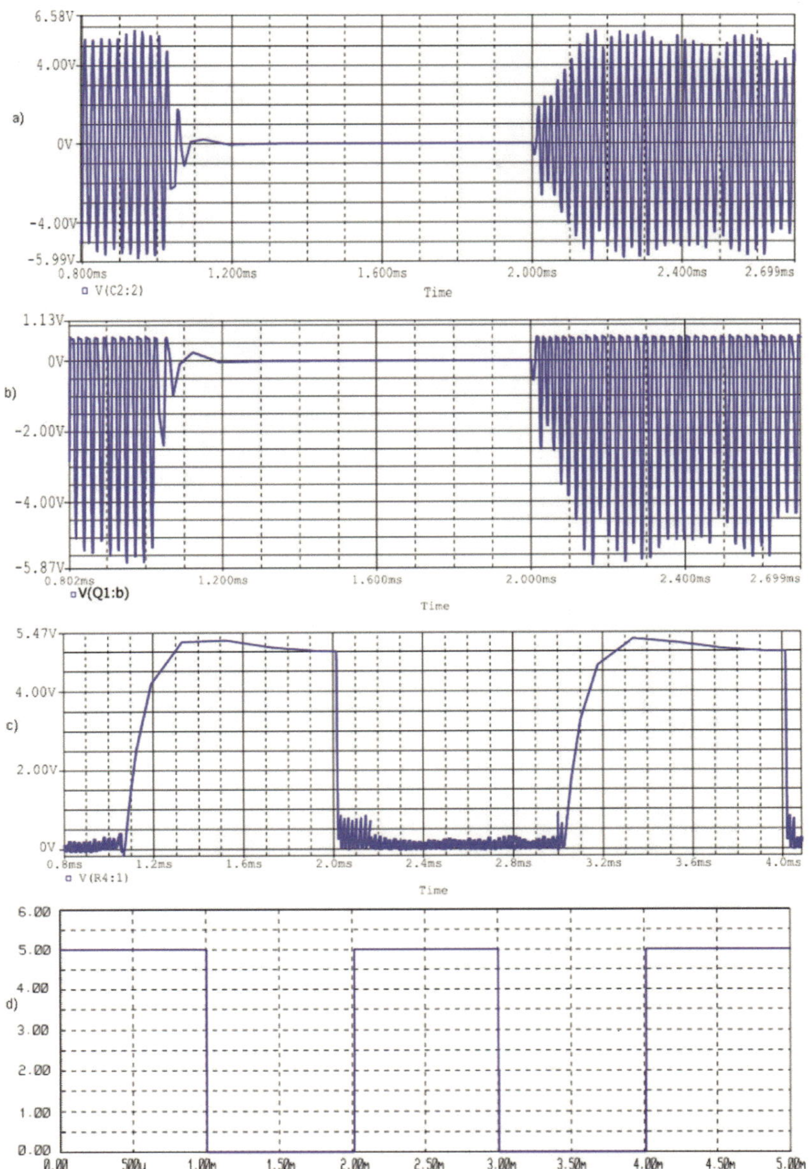

Figure 14. Simulation results at the receiver: (a) signal curve at the low band filtering circuit R1, C1, (b) input signal at the base of transistor Q1, (c) amplified signal at the collector of transistor Q1, (d) output signal A/D conversion.

The configuration of the experimental device is analogous to the circuit shown in **Figure 12** while the transmitter is controlled by a microcontroller and the output of the receive circuit is connected to a microcontroller for signal detection. In order to have a simple operating environment, a 400VDC net power communication system with a single resistive 10 Ohm bus is developed. The experimental results are shown in **Figures 15** and **16**.

In **Figure 16(a)**, the 400VDC net power signal is generated by a microcontroller with the proposed signal transmitter. The original data is modulated by the transmitter and transmitted to

http://dx.doi.org/10.5772/intechopen.78346

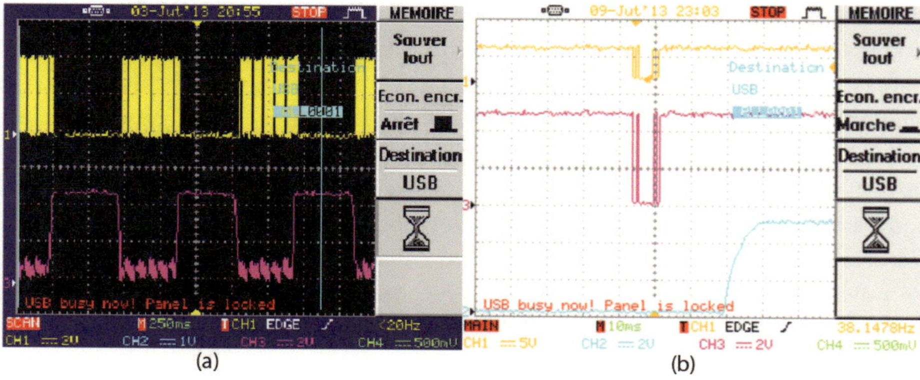

Figure 15. (a) Oscilloscope curves in model tests: yellow modulated signal with microcontroller of the transmitter and violet output signal of the receiver; (b) RS232 low data signal, positive logic: yellow TX/RC6 signal and purple RX/RC7 microcontroller signal.

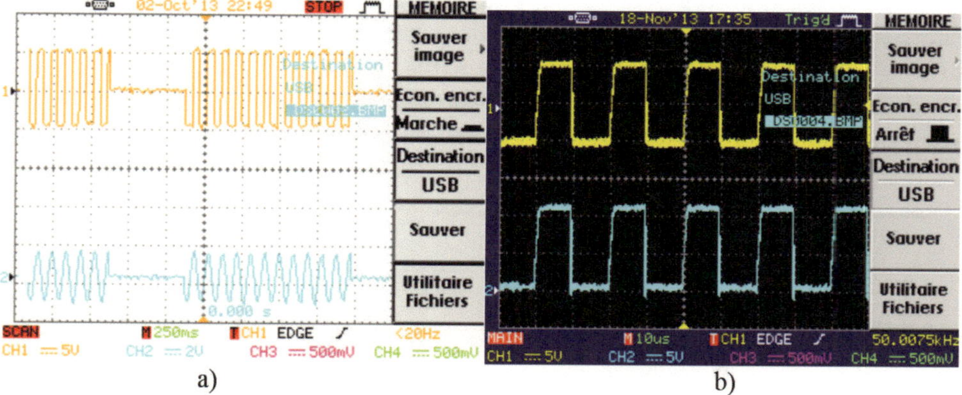

Figure 16. Curves obtained at the oscilloscope: (a) yellow signal curve transmitted on the line and blue output signal curve on the receiver filtering stage; (b) young Lp2 inductance signal and emitter drain signal M1.

the power supply in the form of voltage ripple, it can be seen that the original data is successfully regenerated by the receiver. The receiver output the demodulated signal after the whole packet is received with some delay. In the figures, we find that the experimental results show a good agreement with the initial forecast.

A power line communication system design for 400 V DC net power is presented in this section. The system is essentially a combination of communication system and power distribution system when the communication means is integrated into the power cables. In order to simplify the wiring structure and minimize the amount of system cables, the system structure is designed to be a bus network that the main power cable is pre-installed in the HVDC bus. The simulation results show that the data communication on a low impedance DC net power is achievable with the carrier signal transmitter and the proposed receiver. The main circuit of the transmitter is essentially a split-second converter with the input and output terminal connected together and operates in switching mode to achieve a high signal transmission efficiency. Being controlled by a microcontroller, the carrier transmitter is able to amplitude modulate and transmit a 1 kbps data stream to the power wire. Because the carrier signal travels through the power supply in the form of a ripple current that is filtered by the HVDC bus filter capacitors, to facilitate signal reception, the receiver is designed to be a coupler. Current that obtains the carrier signal from the current of the filter capacitors. The receiver behaves like a simple voltage converter current that converts the signal into sinusoidal coupled current pulses, where the period of the pulse sequence carries the fundamental frequencies of the modulation amplitudes. Signal demodulation and decoding processes are performed by the microcontroller with the program in the code. Practical implementation of the proposed DC net power communication is discussed. The simplicity of the proposed DC power net communication system, it is not only suitable for HVDC bus.

The approach chosen for the renewable energy generator, we have developed a PLC HVDC system based on two types of circuits for a slave circuit interface PLC dedicated to the individual optimizers, 2 a master PLC circuit, that is, the interface with the central management controller. Such a concept is shown in the **Figure 17**.

As a result of these prototypes, the experimental results were obtained even when the transmission module is plugged into the 150 m mains socket from the receiving module, **Figures 18** and **19**. The identification codes of the device sent on the line are correctly received and identified. by their slaves and corresponding devices. However, the circuit shows pick up stray electronic noise in the transmitter-receiver interface circuit of the PLC

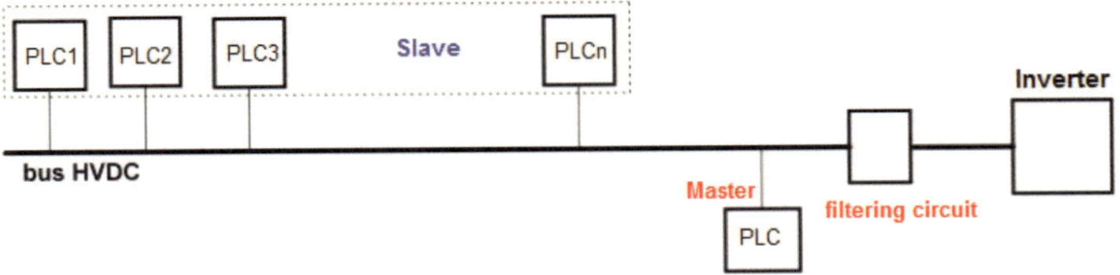

Figure 17. Diagram of master–slave HVDC PLC systems.

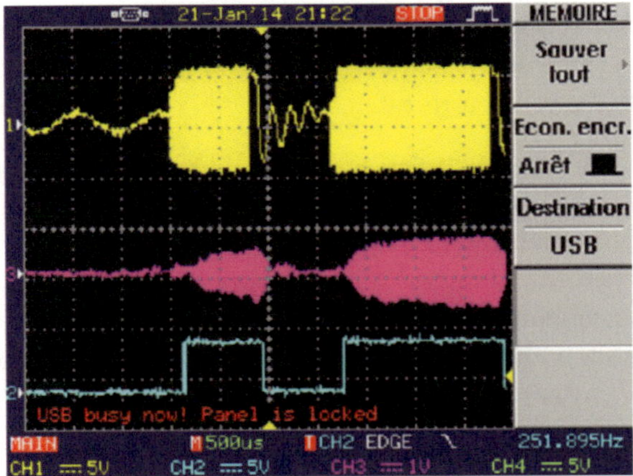

Figure 18. Slave PLC 1 (green signal) and 2 (blue signal) receiver data.

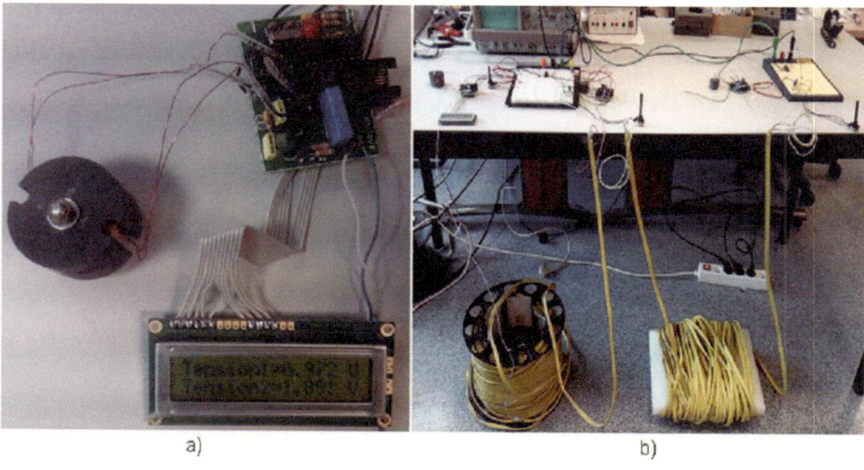

Figure 19. The photos of the experimental in laboratory: (a) master CPL measures the voltages of the two slave PLCs; (b) experimental master PLC system-2 slaves on the cable 100 m.

on the HVDC bus. Thus, in the work in progress, the filter circuits must be modified to have high selectivity, which can be improved by increasing the order of the filter or replacing the passive with an active filter. The problem we are discussing has reeling cores self from PLC master of system.

4. Conclusion

In this chapter we have presented the new developments of dedicated communication system for distributed renewable energy generators. The work involved starts from the definition of constraints to be considered for this specific application to the full realization and testing of prototypes in real conditions. This electronic development is based on the hardware and software implementation of the slave and master modules of a power line communication (PLC) communication systems using the high voltage DC bus connecting all the converters of each DC-DC of the energy generator. The communication protocol used here is the widely accepted Modbus protocol. The communication has been successfully tested and is able to receive and transmit data without error on an experimental long bus. The system works as expected and has been tested to be showing a good response in a noise free environment. After this first realization proving the validity of our choices, work in progress focuses on the improvement of filtering circuits to increase the signal to noise ratio allowing a better selectivity in the information transmitted and detected.

Acknowledgements

This work has been funded by: Quang Ninh University of Industry, Quang Ninh, Vietnam Electric power University, Ha noi, Vietnam Dong Thap University, Dong Thap, Vietnam and Institute of energy sciene, Ha noi, Vietnam.

Author details

Nguyen The Vinh[1*], Vo Thanh Vinh[2], Pham Manh Hai[3], Le Thanh Doanh[3] and Pham Van Duy[4]

*Address all correspondence to: vinhnt@qui.edu.vn

1 Quang Ninh University of Industry, Quang Ninh, Vietnam

2 Dong Thap University, Dong Thap, Vietnam

3 Electric Power University, Ha noi, Vietnam

4 Institute of Energy Science, Ha noi, Vietnam

References

[1] Alonso C. Contribution à l'optimisation, la gestion et le traitement de l'énergie. In: Micro et nanotechnologies/Microélectronique. Toulouse III: Université Paul Sabatier; 2003

[2] Estibals B. Contribution à l'amélioration des chaines de conversion photovoltaiques par l'introduction d'architectures distribuées. In: Micro et nanotechnologies/Microélectronique. Toulouse III: Université Paul Sabatier; 2010

[3] Petibon S. Nouvelles architectures distribuées de gestion et conversion de l'énergie pour les applications photovoltaïques. In: Micro et nanotechnologies/Microélectronique. Toulouse III, Français: Université Paul Sabatier; 2009

[4] Cabal C. Optimisation énergétique de l'étage d'adaptation électronique dédié à la conversion photovoltaïque. In: Micro et nanotechnologies/Microélectronique. Toulouse III, Français: Université Paul Sabatier; 2008

[5] Aelterman P, Rabaey K, Pham HT, Boon N, Verstraete W. Continuous electricity generation at high voltages and currents using stacked microbial fuel cells. Environmental Science & Technology; 2006;**40**(10):3388-3394

[6] Vighetti S. Systèmes photovoltaïques raccordés au réseau: Choix et dimensionnement des étages de conversion. In: Sciences de l'ingénieur [Physics]. Français: Institut National Polytechnique de Grenoble- INPG; 2010

[7] Kolar JW, Friedli T, Krismer F, Looser A, Schweizer M, Steimer P, Bevirt J. Conceptualization and Multi-Objective Optimization of the Electric System of an Airborne Wind Turbine. Swiss Federal Institute of Technology (ETH) Zurich Power Electronic Systems Laboratory. IEEE JESTPE; 2011;**1**(2):73-103

[8] Linares L, Erickson RW, MacAlpine S, Brandemuehl M. Improved energy capture in series string photovoltaics via smart distributed power electronics. Twenty-Fourth Annual IEEE Applied Power Electronics Conference and Exposition, 2009 (APEC 2009); 15-19 February 2009; pp. 904-910, DOI: 10.1109/APEC.2009.4802770

[9] Shimizu T, Hirakata M, Kamezawa T, Watanabe H. Generation control circuit for photovoltaic modules. IEEE Transactions on Power Electronics. 2001;**16**(3):293-300. Publisher Item Identifier S 0885-8993(01)04043-1

[10] Ferrieux JP, Forest F. Alimentation à Découpage, Convertisseurs à Résonance. 3ème ed. Dunod; 2006. ISBN: 2 10 0050539 4

[11] Siri K, Lee CQ, Wu TF. Current distribution control for parallel connected converters: part II. IEEE Trans. Aerospace and Electronic Systems; 1992;**28**:841-851

[12] Louvrier Y. Etude et optimisation multicritères d'un convertisseur DC/DC à canaux multiples entrelacés [Thesis]. Thèse de l'Ecole Polytechnique Fédérale de Lausanne; Thèse No 4718; 2010

[13] Pinomaa A, Ahola J, Kosonen A. Power-line communication-based network architecture for LVDC distribution system. In: IEEE International Symposium on Power-line Communications and Its Applications; May 2011. pp. 358-363. DOI: 10.1109/ISPLC.2011.5764422

[14] Binduhewa PJ, Renfrew AC, Barnes M. MicroGrid power electronic interface for photovoltaics. In: Proceedings of IET Conference on Power Electronics, Machines and Drives; April 2008. ISSN: 0537-9989. DOI: 10.1049/cp:20080523

[15] Bose A. Smart transmission grid applications and their supporting infrastructure. IEEE Transactions on Smart Grid. 2010;**1**(1):11-19. DOI: 10.1109/TSG.2010.2044899. ISSN: 1949-3053

[16] Petit P, Zegaoui A, Sawicki JP, Aillerie M, Charles JP. New architecture for high effeciency DC-DC converter dedicated to photovoltaic conversion. MEDGREEN, Mediterranean, Beyrouth-Liban. 2011;**6**:688-694. DOI: 10.1016/j.egypro.2011.05.078

[17] Kumar D, Rout D, Kumari A, Kumar M, Kumari A, Shamim Ansari MD. Review on automation and bidirectional data transfer based on PLC. Undergraduate Academic Research Journal (UARJ). 2012;**1**(3,4):103-109. ISSN: 2278 – 1129

[18] Vanfretti L, Hertem DV, Nordstrom L, Gjerde JO. A smart transmission grid for Europe: Research challenges in developing grid enabling technologies. In: Power and Energy Society General Meeting, 2011; IEEE; 10 October, 2011. DOI: 10.1109/PES.2011.6039693

[19] Neha S, Ajay T. Design of digital modulators: BASK, BPSK and BFSK using VHDL. International Journal of Advanced Research in Computer Science and Software Engineering. 2013;**3**(1):382-386. ISSN: 2277 128X

[20] Bausch J et al. Characteristics of indoor power line channels in the frequency range 50-500 kHz. In: IEEE International Symposium on Power Line Communications and its Applications; March 26-29; Orlando, Florida, USA; 2006. DOI: 10.1109/ISPLC.2006.247442